SpringerBriefs in Applied Sciences and Technology

Mathematical Methods

Series Editors

Anna Marciniak-Czochra, Institute of Applied Mathematics, IWR, University of Heidelberg, Heidelberg, Germany

Thomas Reichelt, Emmy-Noether research group, Universität Heidelberg, Heidelberg, Germany

More information about this subseries at http://www.springer.com/subseries/11219

Song Wang

The Fitted Finite Volume and Power Penalty Methods for Option Pricing

Springer

Song Wang
School of Electrical Engineering
Computing and Mathematical Sciences
Curtin University
Perth, WA, Australia

Shenzhen Audencia Business School
Shenzhen University
Shenzhen, China

ISSN 2191-530X   ISSN 2191-5318   (electronic)
SpringerBriefs in Applied Sciences and Technology
ISSN 2365-0826   ISSN 2365-0834   (electronic)
SpringerBriefs in Mathematical Methods
ISBN 978-981-15-9557-8   ISBN 978-981-15-9558-5   (eBook)
https://doi.org/10.1007/978-981-15-9558-5

Mathematics Subject Classification: 65N08, 65K15, 91-08, 65M12, 91G20

This Springer imprint is published by the registered company Springer Nature Singapore Pte Ltd.
The registered company address is: 152 Beach Road, #21-01/04 Gateway East, Singapore 189721, Singapore

Preface

A financial derivative is a contract between two parties whose value is dependent on (or derived from) an underlying asset or assets. Derivative securities consist of three major parts: Forwards and Future (obligation to buy or sell), Options (right to buy or sell) and Swaps (simultaneous selling and purchasing). In particular, the first two form the basis of derivative securities.

Options are often used for hedging risks in the underlying assets or stocks. An option can be traded on a secondary financial market. Thus, how to price options accurately has attracted much attention in the past few decades from both financial engineers and researchers. There are mainly two types of options: European option and American option. The former can be exercised only on maturity, while the latter is exercisable anytime prior to or on maturity. Mathematically, the values of European and American options are governed, respectively, by a partial differential equation (PDE), known as Black-Scholes equation, and a differential Linear Complementarity Problem (LCP). Thus, pricing options involve solutions of PDEs and LCPs.

In recent years, we have developed efficient and accurate discretization and optimization methods for numerically solving the aforementioned PDEs and LCPs in both 1- and 2-dimensions. These methods include a fitted finite volume method for the PDEs and a power penalty approach to the LCPs. This book is to put together some of the methods, algorithms, and mathematical analyses developed by us to provide a reference for both practitioners and researchers on the latest advances in numerical methods for pricing financial options. The book also provides materials which can be used in an advanced course on numerical methods in financial engineering for postgraduate research students.

Perth, Australia/Shenzhen, China
July 2020

Song Wang

Contents

Chapter 1
European Options on One Asset

Abstract In this chapter we first give a brief account of stochastic differential equations governing risky asset/stock dynamics and Itô lemma to be used for deducing the mathematical model of pricing European options on one asset. We then derive the Black–Scholes (BS) equation using the ideal of Δ-hedging and Itô's lemma. The BS equation is formulated as a variational problem, which is shown to be uniquely solvable. A fitted Finite Volume Method (FVM) is proposed for the discretization of the equation. We prove that the FVM is unconditionally stable and its solution converges to that of the BS equation. Numerical results are presented to demonstrate the usefulness and accuracy of this FVM.

Keywords European option valuation · Black–Scholes equation · Fitted finite volume method · Stability and convergence.

1.1 Stock Price Dynamics and Itô Lemma

In stochastic mathematics, the Wiener process, denote as W_t, is a stochastic process satisfying the following properties.

1. $W_0 = 0$ and $\Delta W_t := W_{t+\Delta t} - W_t = \varepsilon \sqrt{\Delta t}$ in a small positive time-increment Δt at time t, where ε has a standardized normal distribution $\mathcal{N}(0, 1)$.
2. Increments ΔW_t and ΔW_s in any two different short time intervals are independent.
3. The path $\{W_t : t \geq 0\}$ is continuous in t.

The Wiener's process is the mathematical representation of the 1-dimensional Brownian motion. A stochastic process S_t is said to follow a Geometric Brownian Motion if S_t satisfies the following stochastic differential equation:

$$dS_t = \mu S_t dt + \sigma S_t dW_t, \tag{1.1.1}$$

© The Author(s), under exclusive license to Springer Nature Singapore Pte Ltd. 2020
S. Wang, *The Fitted Finite Volume and Power Penalty Methods for Option Pricing*, SpringerBriefs in Mathematical Methods,
https://doi.org/10.1007/978-981-15-9558-5_1

where μ represents the percentage drift of the process and σ is the percentage volatility (or standard deviation) of the uncertainty in the process. Both μ and σ are constants. In mathematical finance, the price of a risky asset is usually assumed to follow a Geometric Brownian Motion and thus satisfies (1.1.1) (e.g., [8]), which is now called Merton–Black–Scholes model for risky asset price dynamics.

An option on a risky asset is a function of time t and the underlying asset price S satisfying (1.1.1). Itô established in [11] a mathematical presentation of the differential of a time-dependent function of a stochastic process, which is given in the following theorem.

Theorem 1.1.1 (Itô Lemma) *Consider a function $f(t, X)$, where t is time and $X(t)$ satisfies the following stochastic differential equation:*

$$dX = a(t, X)dt + b(t, X)dW, \qquad (1.1.2)$$

with W a Wiener process. If f is a twice-differentiable scalar function, then we have

$$df = \left(\frac{\partial f}{\partial t} + a(t, x) \frac{\partial f}{\partial x} + \frac{1}{2} b^2(t, x) \frac{\partial^2 f}{\partial x^2} \right) dt + b(t, x) \frac{\partial f}{\partial x} dW. \qquad (1.1.3)$$

Equation (1.1.1) is a special case of (1.1.2). If the value of an option. denoted as $V(S, t)$, on a stock whose price S satisfies (1.1.1), then from (1.1.3) we have

$$dV = \left(\frac{\partial f}{\partial t} + \mu S \frac{\partial V}{\partial S} + \frac{1}{2} \sigma^2 S^2 \frac{\partial^2 f}{\partial S^2} \right) dt + \sigma S \frac{\partial f}{\partial S} dW. \qquad (1.1.4)$$

1.2   The Black–Scholes Equation and Its Solvability

In this section we establish the Black–Scholes PDE model for pricing European options, We then reformulate it as a variational problem and show the variational problem is uniquely solvable.

1.2.1   The Black–Scholes Equation

An option is a contract which gives to its owner the right to buy (*call option*) or sell (*put option*) a fixed quantity of a specified asset(s) at a fixed price (*exercise or strike price*) on (European option) or before (American option) a given date (*expiry date*). An option can usually be traded in a financial market and the market prices of the rights to buy and to sell are called *call prices* and *put prices*, respectively. It was shown by Black and Scholes [5] that the price of a European option satisfies a second-order parabolic partial differential equation (PDE), which is now known as the Black–Scholes equation. The Black–Scholes equation can be derived in various ways. In what follows, we derive it using the idea of delta-hedging (see [13]).

Assume that the market conditions are ideal for a European option and its underlying stock, i.e., (i) the short-term interest rate is constant and known, (ii) the stock price follows the geometric Brownian motion (1.1.1); (iii) the stock pays no dividends or distribution; (iv) there are no transaction costs in buying and selling the stock or option. Under these conditions, let us consider the hedging problem: for every European call option issued by you on a stock, how many shares of the stock you need to hold in order to neutralize the risk of the option issued. Mathematically, let us consider a riskless portfolio Π consisting of one European option shorted and δ shares of the underlying stock. At time t, the value of the portfolio is $\Pi = \delta S(t) - V(S(t), t)$, where S denotes the price of the stock and V the value of the option. The differential of V satisfies (1.1.4). In any infinitesimal time interval $[t, t + dt)$, the change in Π is the combination of the changes in S and V, i.e., $d\Pi = \delta dS - dV$. Therefore, combining this relation with (1.1.1) and (1.1.4) we have

$$d\Pi = \left[\mu S\delta - \left(\frac{\partial V}{\partial t} + \frac{1}{2}\sigma^2 S^2 \frac{\partial^2 V}{\partial S^2} + \mu S \frac{\partial V}{\partial S}\right)\right] dt + \sigma S \left(\delta - \frac{\partial V}{\partial S}\right) dW.$$

Since we expect that the portfolio is riskless, the coefficient of dW should vanish, implying that $\delta = \frac{\partial V}{\partial S}$. Thus, the above equality becomes

$$d\Pi = -\left(\frac{\partial V}{\partial t} + \frac{1}{2}\sigma^2 S^2 \frac{\partial^2 V}{\partial S^2}\right) dt. \tag{1.2.1}$$

We expect that this portfolio has the riskless return rate r, i.e., $d\Pi = r\Pi dt = r(S\frac{\partial V}{\partial S} - V)dt$. Comparing this equality with (1.2.1) gives

$$-\left(\frac{\partial V}{\partial t} + \frac{1}{2}\sigma^2 S^2 \frac{\partial^2 V}{\partial S^2}\right) = r\left(S\frac{\partial V}{\partial S} - V\right).$$

Thus, from this equation we have

$$\frac{\partial V}{\partial t} + \frac{1}{2}\sigma^2 S^2 \frac{\partial^2 V}{\partial S^2} + rS\frac{\partial V}{\partial S} - rV = 0. \tag{1.2.2}$$

This is Black–Scholes equation which determines the value of the option. Note that $\frac{\partial V}{\partial S}$ is called the Delta of the option, and is denoted as Δ. The portfolio Π is said to be Δ-neutral, i.e., we long (buy) $\frac{\partial V}{\partial S}$ shares for every option sold, or equivalently long one share and short (sell) $(\frac{\partial V}{\partial S})^{-1}$ options, as in [5].

1.2.2   The Strong Problem

As derived in Sect. 1.2.1, the value V of a European option issued on an asset with price S satisfying (1.1.1) satisfies (1.2.2). We now consider the following generalized Black–Scholes equation (e.g., [22]):

$$LV := -\frac{\partial V}{\partial t} - \frac{1}{2}\sigma^2(t)S^2\frac{\partial^2 V}{\partial S^2} - (r(t)S - D(S,t))\frac{\partial V}{\partial S} + r(t)V = 0, \quad (1.2.3)$$

for $(S, t) \in I \times [0, T)$, with the boundary and terminal (or payoff) conditions

$$V(0, t) = g_1(t), \ V(S_{\max}, t) = g_2(t), \ V(S, T) = g_3(S), \ t \in [0, T), S \in \bar{I}, \tag{1.2.4}$$

where $I = (0, S_{\max}) \subset \mathbb{R}$ with $S_{\max}$ a positive constant greater than the strike price K of the option, and $T > 0$ the expiry date. Note that (1.2.2) is defined on $(0, \infty) \times [0, T)$. However, in (1.2.3)–(1.2.4), we truncate it using $S_{\max}$, in order to be able to solve this problem numerically. We assume that g_1, g_2 and g_3 satisfy the following compatibility conditions

$$g_3(0) = g_1(T) \quad \text{and} \quad g_3(S_{\max}) = g_2(T). \tag{1.2.5}$$

For simplicity, we assume $D(S, t) = d(S, t)S$, where $d(S, t)$ represents the dividend rate and is continuously differentiable. When, $d(S, t) = d(t)$, the problem is said to be path-independent. Otherwise, it is path-independent. There are various choices of final/payoff conditions depending on models. For example, for a call option, the most common final condition is the following payoff function

$$V(x, T) = \max(0, S - K), \quad S \in \bar{I} \tag{1.2.6}$$

where $K < S_{\max}$ denotes the strike/exercise price of the option. A second choice is the *cash-or-nothing* payoff given by

$$V(S, T) = B\mathscr{H}(S - K), \quad S \in \bar{I}, \tag{1.2.7}$$

where $B > 0$ is a constant and $\mathscr{H}$ denotes the Heaviside function. Obviously, this final condition is a step function which is zero if $S < K$ and $S_{\max}$ if $S \geq K$. Another choice is the *bullish vertical spread* payoff defined by

$$V(S, T) = \max(0, S - K_1) - \max(0, S - K_2), \quad x \in \bar{I}, \tag{1.2.8}$$

where K_1 and K_2 are two strike prices satisfying $K_1 < K_2$. This represents a portfolio of buying one call option with the exercise price K_1 and writing another call option with the same expiry date but a larger exercise price (i.e., K_2). For a detailed discussion on this, we refer to [13, 22].

Boundary conditions $g_1(t)$ and $g_2(t)$ are usually difficult to determine exactly when both d and r are non-constant and $S_{\max}$ is finite. The simplest way to determine them for a call option is to choose $V(0, t) = 0$ and $V(S_{\max}, t) = V(S_{\max}, T)e^{r(t-T)}$. Boundary conditions for a European call are determined in [9, Eq. (14)] when both d and r are constant, which can be extended to non-constant d and r as follows

$$V(0, t) = 0, \tag{1.2.9}$$

$$V(S_{\max}, t) = S_{\max} \exp\left(-\int_t^T d(S_{\max}, \tau)d\tau\right) - K \exp\left(-\int_t^T r(\tau)d\tau\right). \tag{1.2.10}$$

Payoff and boundary conditions for put options can be defined analogously.

Before further discussion, we first transform (1.2.3) with the non-homogeneous Dirichlet boundary conditions in (1.2.4) into one with the homogeneous boundary conditions. This can be achieved by adding the term $f(S, t) := -e^{\beta t}LV_0$ to both sides of (1.2.3) and introducing a new variable $u = e^{\beta t}(V(S, t) - V_0(S, t))$, where $\beta > 0$ is a constant to be defined,

$$V_0(S, t) = g_1(t) + \frac{g_2(t) - g_1(t)}{S_{\max}} S \tag{1.2.11}$$

and L is the Black–Scholes differential operator defined in (1.2.3). Under this transformation, (1.2.3) becomes $Lu + \beta u = f$, where L is the operator defined in (1.2.3). This equation can further be written in the following form

$$\mathscr{L}u := -\frac{\partial u}{\partial t} - \frac{\partial}{\partial S}\left[a(t)S^2\frac{\partial u}{\partial S} + b(S, t)Su\right] + c(S, t)u = f(S, t), \tag{1.2.12}$$

where

$$a(t) = \frac{1}{2}\sigma^2(t), \quad b(S, t) = r(t) - d(S, t) - \sigma^2(t), \tag{1.2.13}$$

$$c(S, t) = r(t) + b(S, t) - S\frac{\partial d}{\partial S} + \beta = 2r(t) - \sigma^2(t) - \frac{\partial D(S, t)}{\partial S} + \beta. \tag{1.2.14}$$

From (1.2.4) and the definition of u we see that the boundary and final conditions for (1.2.12) now become

$$u(0, t) = 0 = u(S_{\max}, t), \ t \in [0, T), \ u(S, T) = e^{\beta T}(g_3(S) - V_0(S, T)), \ x \in \bar{I}. \tag{1.2.15}$$

It is also easy to see from (1.2.11) and (1.2.5) that the boundary and final conditions in (1.2.15) satisfy the compatibility conditions since $g_3(0) - V_0(0, T) = 0 = g_3(X) - V_0(X, T)$.

1.2.3   The Variational Problem and Its Solvability

We shall reformulate (1.2.12)–(1.2.15) as a variational problem. To achieve this, it is necessary to introduce some function spaces and norms on them.

For an open interval $\Omega \in \mathbb{R}$, we let $C^m(\Omega)$ (respectively, $C^m(\bar{\Omega})$) be the set of functions, in which a function and its derivatives up to order m are continuous on Ω (respectively $\bar{\Omega}$) for a non-negative integer m. For $1 \le p < \infty$, we let $L^p(\Omega) = \{v : \left(\int_\Omega |v|^p d\Omega\right)^{1/p} < \infty\}$ denote the space of all p-power integrable functions on Ω. For any $p, q \ge 1$ satisfying $1/p + 1/q = 1$, we let $(v, w) := \int_\Omega vw\,d\Omega$ be the duality between $L^p(\Omega)$ and $L^q(\Omega)$, which becomes an inner product when $p = q$. The inner product on $L^2(\Omega)$ is also denoted by $(\cdot, \cdot)$. We use $\| \cdot \|_{L^p(\Omega)}$ to denote the norm on $L^p(\Omega)$. For $m = 1, 2, \ldots$, we let $H^{m,p}(\Omega)$ denote the usual Sobolev space with the norm $\| \cdot \|_{m,p,\Omega}$. When $p = 2$, we simply denote $H^{m,2}(\Omega)$ and $\| \cdot \|_{m,2,\Omega}$ as $H^m(\Omega)$ and $\| \cdot \|_{m,\Omega}$, respectively. When $\Omega = I$, we omit the subscript Ω in the above notation. We put $H_0^m(I) = \{v \in H^m(I) : v(0) = v(S_{\max}) = 0\}$.

To handle the degeneracy in the Black–Scholes equation, we introduce the following weighted L^2-norm $\|v\|_{0,w} := \left(\int_0^{S_{\max}} S^2 v^2 dS\right)^{1/2}$. The space of all weighted square-integrable functions is defined as $L_w^2(I) := \{v : \|v\|_{0,w} < \infty\}$. We also define a weighted inner product on $L_w^2(I)$ by $(u, v)_w := \int_0^{S_{\max}} S^2 uv\,dS$.

Using a standard argument (cf., for example, [6, Chaps. 1 & 2]), it is easy to show that the pair $(L_w^2(I), (\cdot, \cdot)_w)$ is a Hilbert space. For brevity, we omit this discussion. Using $L^2(I)$ and $L_w^2(I)$, we define the following weighted Sobolev space

$$H_{0,w}^1(I) := \{v : v \in L^2(I), v' \in L_w^2(I) \text{ and } v(S_{\max}) = 0\}.$$

Let $\| \cdot \|_{1,w}$ be a functional on $H_{0,w}^1(I)$ defined by

$$\|v\|_{1,w} = (\|v\|_0^2 + \|v'\|_{0,w}^2)^{1/2} = \left[(S^2 v', v') + (v, v)\right]^{1/2}. \tag{1.2.16}$$

Then, it is easy to check that $\| \cdot \|_{1,w}$ is a weighted H^1-norm (energy norm) on $H_{0,w}^1(I)$. Using the inner products on $L^2(I)$ and $L_w^2(I)$, we define a weighted inner product on $H_{0,w}^1(I)$ by $(\cdot, \cdot)_H := (\cdot, \cdot) + (\cdot, \cdot)_w$. For the pair $(H_{0,w}^1(I), (\cdot, \cdot)_H)$, we have the following lemma.

Lemma 1.2.1 *The pair $(H_{0,w}^1(I), (\cdot, \cdot)_H)$ is a Hilbert space.*

The result is obvious since both pairs $(L^2(I), (\cdot, \cdot))$ and $(L_w^2(I), (\cdot, \cdot)_w)$ are Hilbert spaces. For brevity, we omit a formal proof of this lemma, but refer the reader to a similar proof in [6, Chap. 2]. For a detailed discussion of weighted Sobolev spaces, we refer to [14].

Remark 1.2.1 It is easy to check by examples that $H_{0,w}^1(I)$ contains the conventional Sobolev space $H_0^1(I)$ as a proper subspace. Also, for a $v \in H_{0,w}(I)$, $v' \in L_w^2(I)$, i.e., $\int_0^{S_{\max}} S^2 (v')^2 dS < \infty$. Intuitively, this implies that $(Sv')^2 \sim S^q$ for some $q > -1$

when S is close to 0. From this we see that near 0, $Sv' \sim S^{q/2}$, and so $Sv \sim S^{1+q/2}$ for some $q > -1$. Therefore, if $v, w \in H_{0,w}(I)$, $S^2 v' w \sim S^{1-q}$ near $S = 0$ some $q > -1$. This further implies that $\lim_{S \to 0^+} S^2 v' w = 0$. This intuition will be used in the formulation of the variational problem corresponding to (1.2.12)–(1.2.15).

We now show the variational problem corresponding to (1.2.12)–(1.2.15) has a unique solution in $H^1_{0,w}(I)$ for $t \in [0, T]$ almost everywhere (a.e.). We will often write $u(\cdot, t)$ as $u(t)$ when we regard it as an element of $H^1_{0,w}(I)$. From time to time, we will also suppress the independent time variable t (or τ) when doing so causes no confusion.

Before further discussion, make the the following assumptions:

Assumption 1.2.1

1. The functions $\sigma(t)$ and $\frac{\partial D(S,t)}{\partial S}$ are continuous and there exist positive constants $\overline{\sigma}, \underline{\sigma}, \overline{r}$ and $\overline{d}$ such that $\underline{\sigma} \le \sigma(t) \le \overline{\sigma}, 0 \le r(t) \le \overline{r}$ and $\left| \frac{\partial D}{\partial S} \right| \le \overline{d}$.

2. The constant β is chosen such that $\beta - \sup_{(x,t) \in I \times (0,T)} \left(\sigma^2(t) + \frac{\partial D(S,t)}{\partial S} \right) \ge \beta_0$, where β_0 is a positive constant.

Now, for any $v \in H^1_{0,w}(I)$, multiplying both sides of (1.2.12) by v, integrating the resulting equation on I and using integration by parts we have

$$\left(-\frac{\partial u}{\partial t}, v \right) + \left(aS^2 \frac{\partial u}{\partial S} + bSu, \frac{\partial v}{\partial S} \right) + (cu, v) = (f, v)$$

for any $t \in [(0, T)$, since $S^2 \frac{\partial u}{\partial S} v \to 0$ as $S \to 0$ or $S_{\max}$ as discussed in Remark 1.2.1. In the above we used the homogeneous boundary conditions (1.2.15) in the strong form of the problem. This motivates us to pose the following variational problem:

Problem 1.2.1 *Find $u(t) \in H^1_{0,w}(I)$ for $t \in (0, T)$ a.e. such that for all $v \in H^1_{0,w}(I)$*

$$\left(-\frac{\partial u(t)}{\partial t}, v \right) + A(u(t), v; t) = (f(t), v) \tag{1.2.17}$$

and (1.2.15) is satisfied, where $A(\cdot, \cdot; t)$ is the following bilinear form on $H^1_{0,w}(I)$:

$$A(v, w; t) := \left(aS^2 \frac{\partial v}{\partial S} + bSv, \frac{\partial w}{\partial S} \right) + (cv, w), \quad v, w \in H^1_{0,w}(I). \tag{1.2.18}$$

The following theorem shows that Problem 1.2.1 is uniquely solvable under Assumption 1.2.1.

Theorem 1.2.1 *Let Assumption 1.2.1 be satisfied. Then, there exists a unique solution to Problem 1.2.1.*

Proof To prove this theorem, it is sufficient to show that A defined in (1.2.18) is coercive and continuous. Integrating by parts, we have, for any $v \in H^1_{0,w}(I)$,

$$\int_0^{S_{\max}} bSvv'dS = -\int_0^{S_{\max}} \left(b + S\frac{\partial b}{\partial S}\right)v^2dS - \int_0^{S_{\max}} bSvv'dS.$$

From this we have

$$\int_0^{S_{\max}} bSvv'dS = -\frac{1}{2}\int_0^{S_{\max}} \left(b + S\frac{\partial b}{\partial S}\right)v^2dS = -\frac{1}{2}\int_0^{S_{\max}} \left(b - S\frac{\partial b}{\partial S}\right)v^2dS.$$

In the above we used (1.2.13). Substituting the above into the bilinear form on the right-hand side of (1.2.17) and using (1.2.13) and (1.2.14) we have

$$A(v, v; t) = (aS^2v', v') + \left(\left(r + \frac{b}{2} - \frac{S}{2}\frac{\partial d}{\partial S}\right)v, v\right)$$

$$= (aS^2v', v') + \frac{1}{2}\left(\left(3r - \sigma^2 - \frac{\partial D}{\partial S}\right)v, v\right)$$

$$\geq \frac{1}{2}\min\{\underline{\sigma}^2, 3r + \beta_0\}\left[(aS^2v', v') + (v, v)\right] \geq C\|v\|^2_{1,w} \qquad (1.2.19)$$

by Assumption 1.2.1, where C denotes a positive constant, independent of v and $\|\cdot\|_{1,w}$ is defined in (1.2.16). Therefore, $A(\cdot, \cdot; t)$ is coercive on $H^1_{0,w}(I)$.

We now prove that $A(\cdot, \cdot; t)$ is Lipschitz continuous. For any $v, w \in H^1_{0,w}(I)$, using Cauchy–Schwarz inequality we get from (1.2.18)

$$A(v, w; t) \leq M\left(\|v'\|_{0,w}\|w'\|_{0,w} + \|v\|_0\|w'\|_{0,w} + \|v\|_0\|w\|_0\right)$$

$$\leq M(\|v'\|_{0,w} + \|v\|_0)(\|w'\|_{0,w} + \|w\|_0) \leq M\|v\|_{1,w}\|w\|_{1,w},$$

where M denotes a generic positive constant, independent of v and w, but depending on the constants in Assumption 1.2.1. Therefore, A is Lipschitz continuous with respect to the two variables. Furthermore, $H^1_{0,w}(I)$ is a Hilbert space by Lemma 1.2.1. Using an existing theoretical results (cf., for example, [12, Lemma 1 and Theorem 1.33]), we see that Problem 1.2.1 has a unique solution. □

Remark 1.2.2 Assumption 1.2.1(2) is not essential for proving the unique solvability of Problem 1.2.1. In fact, the condition (1.2.19) can be replaced by Gårding inequality [17, Theorem 9.17] $A(v, v; t) \geq C_1\|v\|^2_{1,w} - C_2\|v\|^2_0$ for positive constants C_1 and C_2. Thus, when $\beta = 0$ and $\beta_0 < 0$, Problem 1.2.1 is still uniquely solvable, which is reasonable as the solvability of the problem does not depends on the scaling factor $e^{\beta t}$.

Remark 1.2.3 A special regularity assumption on the continuity of u or, equally, V at the boundary $S = 0$ needs to be made. This is because we cannot conclude that for any $u \in H^1_{0,w}(\Omega)$, u is continuous at $S = 0$. This continuity has to be established separately, see, e.g., [1], where the case with $d = 0$, $r = $ const and $\sigma = \sigma(S, t)$ is considered. We assume that the solution possesses the following property:

$$\exists g_1 : (0, T) \to \mathbb{R} : \lim_{S \to 0} V(S, t) = g_1(t) \quad \forall t \in (0, T). \tag{1.2.20}$$

1.3 The Fitted Finite Volume Method (FVM)

1.3.1 The Formulation of the FVM

There are several existing methods developed for solving (1.2.3) (or (1.2.12)) such as those in [2, 7, 10, 18, 20, 22]. However, most of these methods are unable to handle the degeneracy of $\mathscr{L}$ as $S \to 0^+$. Due to this degeneracy, the solution to (1.2.12) cannot take a 'trace' as $S \to 0^+$. A fitted FVM is developed in [21] to handle the degeneracy at the discrete level, based on the fitting techniques used in [15, 16]. The idea of fitting can be traced back to [3].

Let the interval $I = (0, S_{\max})$ be divided into N sub-intervals $I_i := (S_i, S_{i+1})$, $i = 0, 1, \ldots, N - 1$, with $0 = S_0 < S_1 < \cdots < S_N = S_{\max}$. For each $i = 0, 1, \ldots, N - 1$, we put $h_i = S_{i+1} - S_i$ and $h = \max_{0 \le i \le N-1} h_i$. We also let $S_{i-1/2} = (S_{i-1} + S_i)/2$ and $S_{i+1/2} = (S_i + S_{i+1})/2$ for each $i = 1, 2, \ldots, N - 1$. These mid-points form a second partition of $(0, S_{\max})$ if we define $S_{-1/2} = S_0$ and $S_{N+1/2} = S_N$.

Integrating both sides of (1.2.12) over $(S_{i-1/2}, S_{i+1/2})$ we have

$$-\int_{S_{i-1/2}}^{S_{i+1/2}} \frac{\partial u}{\partial t} dS - \left[S \left(aS \frac{\partial u}{\partial S} + bu \right) \right]_{S_{i-1/2}}^{S_{i+1/2}} + \int_{S_{i-1/2}}^{S_{i+1/2}} cu\, dS = \int_{S_{i-1/2}}^{S_{i+1/2}} f\, dS$$

for $i = 1, 2, \ldots, N - 1$. Applying the mid-point quadrature rule to the first, third and last terms gives

$$-\frac{\partial u_i}{\partial t} l_i - \left[S_{i+1/2} \rho(u)|_{S_{i+1/2}} - S_{i-1/2} \rho(u)|_{S_{i-1/2}} \right] + c_i u_i l_i = f_i l_i \tag{1.3.1}$$

for $i = 1, 2, \ldots, N - 1$, where $l_i = S_{i+1/2} - S_{i-1/2}$, $c_i = c(S_i, t)$, $f_i = f(S_i, t)$, u_i is an approximation to $u(S_i, t)$ and $\rho(u)$ is a flux defined by

$$\rho(u) := aS \frac{\partial u}{\partial S} + bu. \tag{1.3.2}$$

Clearly, we now need to derive approximations to $\rho(u)$ at $S_{i+1/2}$, $i = 0, 1, \ldots, N - 1$. This discussion is divided into two cases with $i \ge 1$ and $i = 0$, respectively.
Case I. Approximation of ρ at $S_{i+1/2}$ for $i \ge 1$.
Let us consider the following two-point boundary value problem:

$$(aSv' + b_{i+1/2}v)' = 0, \quad S \in I_i, \quad v(S_i) = u_i, \quad v(S_{i+1}) = u_{i+1}, \tag{1.3.3}$$

where $b_{i+1/2} = b(S_{i+1/2}, t)$. Integrating the equation in (1.3.3) yields

$$\rho_i(v) := aSv' + b_{i+1/2}v = C_1, \tag{1.3.4}$$

where C_1 denotes an additive constant. The integrating factor of this 1st-order linear equation is $\mu = S^{b_{i+1/2}/a}$ and the analytic solution is

$$v = S^{-b_{i+1/2}/a} \left(\int S^{b_{i+1/2}/a} \frac{C_1}{aS} dS + C_2 \right) = \frac{C_1}{b_{i+1/2}} + C_2 S^{-b_{i+1/2}/a}, \tag{1.3.5}$$

where C_2 is also an additive constant. In the above, we assume that $b_{i+1/2} \neq 0$. This restriction will be remedied later. Applying the boundary conditions in (1.3.3) to (1.3.5) we obtain

$$u_i = \frac{C_1}{b_{i+1/2}} + C_2 S_i^{-\alpha_i}, \quad \text{and} \quad u_{i+1} = \frac{C_1}{b_{i+1/2}} + C_2 S_{i+1}^{-\alpha_i}, \tag{1.3.6}$$

where $\alpha_i = b_{i+1/2}/a$. Solving this linear system gives

$$\rho_i(u) = C_1 = b_{i+1/2} \frac{S_{i+1}^{\alpha_i} u_{i+1} - S_i^{\alpha_i} u_i}{S_{i+1}^{\alpha_i} - S_i^{\alpha_i}}. \tag{1.3.7}$$

This gives a representation for the flux on the right-hand side of (1.3.4). Note that (1.3.7) also holds when $\alpha_i \to 0$. This is because

$$\lim_{\alpha_i \to 0} \frac{S_{i+1}^{\alpha_i} - S_i^{\alpha_i}}{b_{i+1/2}} = \frac{1}{a} \lim_{\alpha_i \to 0} \frac{S_{i+1}^{\alpha_i} - S_i^{\alpha_i}}{\alpha_i} = \frac{1}{a}(\ln S_{i+1} - \ln S_i) > 0, \tag{1.3.8}$$

since $S_i < S_{i+1}$ and $a > 0$. Obviously, $\rho_i(u)$ in (1.3.7) provides an approximation to the flux $\rho(u)$ at $S_{i+1/2}$.

Case II. Approximation of ρ at $S_{1/2}$.

The analysis in Case I does not apply to the approximation of the flux on $(0, S_1)$, because (1.3.3) is degenerate. This can be seen from the expression (1.3.5). When $\alpha_0 > 0$, we have to choose $C_2 = 0$ as, otherwise, v blows up as $S \to 0$. However, the resulting solution $v = C_1/b_{1/2}$ can never satisfy both of the conditions in (1.3.3) unless $u_0 = u_1$. To solve this difficulty, let us re-consider (1.3.3) with an extra degree of freedom in the following form

$$(aSv' + b_{1/2}v)' = C_2, \quad \text{in } (0, S_1) \quad \lim_{S \to 0+} v(S) = u_0, \quad v(S_1) = u_1, \tag{1.3.9}$$

where C_2 is an unknown constant to be determined and $u_0 = g_1(t)$ by (1.2.20). Integrating (1.3.9) once we have $aSv' + b_{1/2}v = C_2S + C_3$. Using the condition

$\lim_{S \to 0+} v(0) = u_0$ we have $C_3 = b_{1/2} u_0$, and so the above equation becomes

$$\rho_0(u) := aSv' + b_{1/2}v = b_{1/2}u_0 + C_2 S. \qquad (1.3.10)$$

Solving this problem analytically gives

$$v = \begin{cases} u_0 + \frac{C_2 S}{a + b_{1/2}} + C_4 S^{-\alpha_0} & \alpha_0 \neq -1, \\ u_0 + \frac{C_2}{a} S \ln x + C_4 S & \alpha_0 = -1, \end{cases} \qquad (1.3.11)$$

where C_4 is an additive constant (depending on t).

To determine the constants C_2 and C_4, we first consider the case that $\alpha_0 \neq -1$. When $\alpha_0 \geq 0$, $v(0) = u_0$ implies that $C_4 = 0$. If $\alpha_0 < 0$, C_4 is arbitrary, so we also choose $C_4 = 0$. Using $v(S_1) = u_1$ in (1.3.9) we obtain $C_2 = \frac{1}{S_1}(a + b_{1/2})(u_1 - u_0)$.

When $\alpha_0 = -1$, from (1.3.11) we see that $v(0) = u_0$ is satisfied for any C_2 and C_4. Therefore, solutions to C_2 and C_4 are not unique. We choose $C_2 = 0$, and $v(S_1) = u_1$ in (1.3.9) gives $C_4 = (u_1 - u_0)/S_1$. Therefore, from (1.3.10) we have that

$$\rho_0(u) = (aSv' + b_{1/2}v)_{S_{1/2}} = \frac{1}{2}[(a + b_{1/2})u_1 - (a - b_{1/2})u_0] \qquad (1.3.12)$$

for both $\alpha_0 = -1$ and $\alpha_0 \neq -1$. Furthermore, (1.3.11) reduces to

$$v = u_0 + (u_1 - u_0)S/S_1 \qquad S \in [0, S_1]. \qquad (1.3.13)$$

Now, using (1.3.7) and (1.3.12) obtained in Case I and Case II above respectively, we define a global piecewise constant approximation to $\rho(u)$ by $\rho_h(u)$ satisfying

$$\rho_h(u) = \rho_i(u) \quad \text{if } S \in I_i. \quad i = 0, 1, \ldots, N - 1. \qquad (1.3.14)$$

Substituting (1.3.7) into (1.3.14) and then the result and (1.3.12) into (1.3.1) we obtain

$$-\frac{\partial u_i}{\partial t} l_i + e_{i,i-1}u_{i-1} + e_{i,i}u_i + e_{i,i+1}u_{i+1} = f_i l_i, \qquad (1.3.15)$$

where

$$e_{1,0} = -\frac{S_1}{4}(a - b_{1/2}), \quad e_{1,2} = -\frac{b_{1+1/2}S_{1+1/2}S_2^{\alpha_1}}{S_2^{\alpha_1} - S_1^{\alpha_1}}, \qquad (1.3.16)$$

$$e_{1,1} = \frac{S_1}{4}(a + b_{1+1/2}) + \frac{b_{1+1/2}S_{1+1/2}S_1^{\alpha_1}}{S_2^{\alpha_1} - S_1^{\alpha_1}} + c_1 l_1, \qquad (1.3.17)$$

and

$$e_{i,i-1} = -\frac{b_{i-1/2}S_{i-1/2}S_{i-1}^{\alpha_{i-1}}}{S_i^{\alpha_{i-1}} - S_{i-1}^{\alpha_{i-1}}}, \quad e_{i,i+1} = -\frac{b_{i+1/2}S_{i+1/2}S_{i+1}^{\alpha_i}}{S_{i+1}^{\alpha_i} - S_i^{\alpha_i}}, \tag{1.3.18}$$

$$e_{i,i} = \frac{b_{i-1/2}S_{i-1/2}S_i^{\alpha_{i-1}}}{S_i^{\alpha_{i-1}} - S_{i-1}^{\alpha_{i-1}}} + \frac{b_{i+1/2}S_{i+1/2}S_i^{\alpha_i}}{S_{i+1}^{\alpha_i} - S_i^{\alpha_i}} + c_i l_i, \tag{1.3.19}$$

for $i = 2, 3, \ldots, N - 1$. These form an $(N - 1) \times (N - 1)$ linear system for $\boldsymbol{u} :=$ $(u_1(t), \ldots, u_N(t))^\top$ with $u_0(t)$ and $u_N(t)$ in (1.3.15) equal to the given homogeneous boundary conditions in (1.2.15).

1.3.2 Time Discretization

We now discuss the time-discretization of the linear ODE system (1.3.15). Let E_i, $i = 1, 2, \ldots, N - 1$ be $1 \times (N - 1)$ row vectors defined by

$$E_1 = (e_{11}(t), e_{12}(t), 0, \ldots, 0), \quad E_{N-1} = (0, \ldots, 0, e_{N-1,N}(t), e_{N-1,N-1}(t)),$$
$$E_i = (0, \ldots, 0, e_{i,i-1}(t), e_{i,i}(t), e_{i,i+1}(t), 0, \ldots, 0), \quad i = 2, 3, \ldots, N - 2,$$

where $e_{i,i-1}$, $e_{i,i}$ and $e_{i,i+1}$ are defined in (1.3.16)–(1.3.19) and those which are not defined are zeros. Obviously, using E_i, (1.3.15) can be rewritten as

$$-\frac{\partial u_i(t)}{\partial t} l_i + E_i(t)\boldsymbol{u}(t) = f_i(t)l_i \tag{1.3.20}$$

for $i = 1, 2, \ldots, N - 1$. This is a first order linear ODE system. To discretize this system, we let t_i $(i = 0, 1, \ldots, K)$ be a set of partition points on $[0, T]$ satisfying $T = t_0 > t_1 > \cdots > t_K = 0$. Then, we apply the two-level implicit time-stepping method with a splitting parameter $\theta \in [1/2, 1]$ to (1.3.20) to yield

$$\frac{u_i^{k+1} - u_i^k}{-\Delta t_k} l_i + \theta E_i^{k+1} \boldsymbol{u}^{k+1} + (1 - \theta) E_i^k \boldsymbol{u}^k = (\theta f_i^{k+1} + (1 - \theta) f_i^k) l_i \tag{1.3.21}$$

for $k = 0, 1, \ldots, K - 1$, where $\Delta t_k = t_{k+1} - t_k < 0$, $E_i^k = E_i(t_k)$, $f_i^k = f(S_i, t_k)$ and $\boldsymbol{u}^k$ denotes the approximation of $\boldsymbol{u}$ at $t = t_k$. Let $E^k = (E_1^k, E_2^k, \ldots, E_{N-1}^k)^\top$. Then, the above linear system can be re-written as

$$(\theta E^{k+1} + G^k)\boldsymbol{u}^{k+1} = \boldsymbol{f}^k + [G^k - (1 - \theta)E^k]\boldsymbol{u}^k, \quad k = 0, 1, \ldots, K - 1, \tag{1.3.22}$$

where $G^k = \mathrm{diag}(l_1/(-\Delta t_k), \ldots, l_{N-1}/(-\Delta t_k))$ and $\boldsymbol{f}^k = \theta(f_1^{k+1}l_1, \ldots, f_{N-1}^{k+1} l_{N-1})^\top + (1 - \theta)(f_1^k l_1, \ldots, f_{N-1}^k l_{N-1})^\top$.

When $\theta = 1/2$, the time-stepping scheme becomes that of the Crank-Nicolson and when $\theta = 1$, it is the backward Euler (or fully implicit) scheme. Both of the two cases are unconditionally stable, and they are of second and first order accuracy, respectively.

We now show that, when $|\Delta t_k|$ is sufficiently small, the system matrix of (1.3.22) is an M-matrix. This is contained in the following theorem.

Theorem 1.3.1 *For any given $k = 1, 2, \ldots, K - 1$, if $|\Delta t_k|$ is sufficiently small, the the system matrix of (1.3.22) is an M-matrix.*

Proof Let us first investigate the off-diagonal entries of E^{k+1} in (1.3.22). From (1.3.16)–(1.3.19) we see that $e_{i,j} \leq 0$ for all $i, j = 1, 2, \ldots, N - 1$, $j \neq i$. This is because $\frac{b_{i+1/2}}{S_{i+1}^{\alpha_i} - S_i^{\alpha_i}} = \frac{a\alpha_i}{S_{i+1}^{\alpha_i} - S_i^{\alpha_i}} > 0$ for all $i = 1, 2, \ldots, N - 1$ and all $b_{i+1/2} \neq 0$. From (1.3.8) we see that this also holds when $b_{i+1/2} \to 0$. This proves that all of the off-diagonal element of the system matrix of (1.3.22) are non-positive.

Furthermore, from (1.3.16)–(1.3.19) and the definitions of E_i^{k+1}, $i = 1, 2, \ldots$, $N - 1$, it is easy to check that the diagonal entries of $(\theta E^{k+1} + G^k)$ are given by

$$\frac{l_1}{-\Delta t_k} + \theta e_{1,1}^{k+1} = \theta \left(\sum_{j=1}^{N-1} |e_{1,j}^{k+1}| \right) + \theta \frac{S_1}{4}(a^{k+1} + b_{1/2}^{k+1}) + \left(\theta c_1^{k+1} + \frac{1}{|\Delta t_k|} \right) l_1,$$

$$\frac{l_j}{-\Delta t_k} + \theta e_{i,i}^{k+1} = \theta \left(\sum_{j=1}^{N-1} |e_{i,j}^{k+1}| \right) + \left(\theta c_j^{k+1} + \frac{1}{|\Delta t_k|} \right) l_j$$

for $i = 2, 3, \ldots, N - 1$. Thus, when $|\Delta t_k|$ is sufficiently small, $\theta E^{k+1} + G^k$ is (strictly) diagonally dominant. Therefore, it is an M-matrix [19, p. 85]. □

Remark 1.3.1 We remark $e_{1,0}$ defined in (1.3.16) may not be negative. However, this will not make any difficulty to the method because $e_{1,0}$ does not appear in the system matrix of (1.3.22).

Remark 1.3.2 Theorem 1.3.1 shows that the fully discretized system (1.3.22) satisfies the discrete maximum principle. This guarantees that numerical solutions from this method are non-negative as the option prices should be.

Finally, we comment that local approximations to $\partial u/\partial S$ and $\partial^2 u/\partial S^2$, can be obtained easily from (1.3.13) and (1.3.5). These two quantities, known respectively as the Δ and Γ of an option, are useful in practice. In particular, the former is used by financial engineers for constructing portfolios that hedge against risk (or portfolios that are *delta neutral*). This is also known as *delta hedging*.

1.3.3 Stability and Convergence

We shall re-formulate the FVM as finite element method and present a stability and convergence analysis for the scheme, based on those in [4, 21].

1.3.3.1 The Finite Element Formulation of the FVM

For any $i = 1, 2, \ldots, N - 1$, let ψ_i be defined by

$$\psi_i = \begin{cases} 1, & S \in I_i := (S_{i-1/2}, S_{i+1/2}), \\ 0, & \text{otherwise,} \end{cases} \tag{1.3.23}$$

We choose the test space to be $V_h = \text{span}\{\psi_i\}_1^{N-1}$.

To define the trial space, we choose the hat function ϕ_i associated with S_i in the following way. On I_i we choose ϕ_i so that it satisfies (1.3.3) with $\phi_i(S_i) = 1$ and $\phi_i(S_{i+1}) = 0$. Naturally, the solution to this two-point boundary value problem is given in (1.3.5) where C_1 and C_2 are determined by (1.3.6) with $u_i = 1$ and $u_{i+1} = 0$. Similarly, we can define $\phi_i(x)$ on I_{i-1} so that $\phi_i(S_{i-1}) = 0$ and $\phi_i(S_i) = 1$. Combining these two solutions, choose

$$\phi_i(S) = \begin{cases} \left[1 - \left(\frac{S_i}{S_{i-1}}\right)^{\alpha_{i-1}}\right]^{-1} \left[1 - \left(\frac{S}{S_{i-1}}\right)^{\alpha_{i-1}}\right], & S \in I_{i-1}, \\ \left[1 - \left(\frac{S_i}{S_{i+1}}\right)^{\alpha_i}\right]^{-1} \left[1 - \left(\frac{S}{S_{i+1}}\right)^{\alpha_i}\right], & S \in \bar{I}_i, \\ 0, & \text{otherwise.} \end{cases} \tag{1.3.24}$$

One $[0, S_1]$, $\phi_1(S)$ is given by the linear function in (1.3.13) with $u_0 = 0$ and $u_1 = 1$. For this set of basis functions we have the following theorem.

Theorem 1.3.2 *For each $i = 1, \ldots, N - 1$, the function ϕ_i is monotonically increasing and decreasing on I_{i-1} and I_i respectively. Furthermore, ϕ_i and ϕ_{i+1} satisfy $\phi_i(S) + \phi_{i+1}(S) = 1$ on I_i for $i = 1, 2, \ldots, N - 1$.*

Proof Differentiating ϕ_i on I_i we have $\phi_i'(S) = \frac{-\alpha_i}{1 - \left(\frac{S_i}{S_{i+1}}\right)^{\alpha_i}} \frac{S^{\alpha_i}}{S_{i+1}^{\alpha_i}}$ on (S_i, S_{i+1}). Since $S_{i+1} - S_i > 0$, we have $\frac{\alpha_i}{1 - (S_i/S_{i+1})^{\alpha_i}} > 0$ for all α_i. Therefore, $\phi'(x) < 0$, and so ϕ is monotonically decreasing in I_i. Similarly, it is easy to show that ϕ_i is monotonically increasing in I_{i-1}.

From (1.3.24) we have, for $S \in I_i$,

$$\phi_i + \phi_{i+1} = \frac{1 - \left(\frac{S}{S_{i+1}}\right)^{\alpha_i}}{1 - \left(\frac{S_i}{S_{i+1}}\right)^{\alpha_i}} + \frac{1 - \left(\frac{S}{S_i}\right)^{\alpha_i}}{1 - \left(\frac{S_{i+1}}{S_i}\right)^{\alpha_i}} = \frac{S_{i+1}^{\alpha_i} - S^{\alpha_i}}{S_{i+1}^{\alpha_i} - S_i^{\alpha_i}} + \frac{S_i^{\alpha_i} - S^{\alpha_i}}{S_i^{\alpha_i} - S_{i+1}^{\alpha_i}} = 1.$$

Thus, we have proved this theorem.   □

Two examples of these hat functions with constant α are plotted in Fig. 1.1.

The finite element trial space is chosen to be $U_h = \text{span}\{\phi_i\}_1^{N-1}$. For an arbitrary function $v \in C(\bar{I})$, we define the mass lumping operator $L_h : C(\bar{I}) \to V_h$ by

$$L_h(v) := \sum_{i=0}^{N} v(S_i)\Psi_i(S), \tag{1.3.25}$$

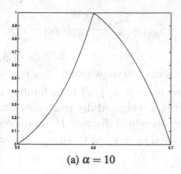 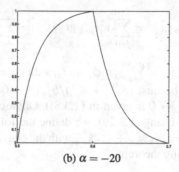

(a) $\alpha = 10$ (b) $\alpha = -20$

Fig. 1.1 Examples of the hat functions for different values of α

where Ψ_i is defined in (1.3.23).

Using U_h, V_h and L_h, we define the following Petrov–Galerkin problem.

Problem 1.3.1 *Find $u_h(t) \in U_h$ such that for all $v_h \in V_h$*

$$- (L_h(\dot{u}), v_h) + A_h(u_h, v_h; t) = (L_h(f), v_h), \qquad (1.3.26)$$

where $A_h(\cdot, \cdot; t)$ is the bilinear form on $U_h \times V_h$ defined by

$$A_h(u_h, v_h; t) := - \sum_{j=1}^{N-1} \left[x\left(a(t)x\frac{\partial u_h(t)}{\partial x} + \hat{b}(t)u_h(t)\right) \right]_{S_{j-1/2}}^{S_{j+1/2}} v_h + (L_h(c(t)u_h(t)), v_h)$$

$$(1.3.27)$$

and $\hat{b}$ is piecewise constant satisfying $\hat{b}(t) = b_{j+1/2}(t)$ when $S \in I_i$ for all feasible i.

From the constructions of the FVM, ϕ_i and ψ_j, it is easy to verify that if $u_h(t) = \sum_{j=1}^{N-1} u_j(t)\phi_j$ and $v_h = \psi_i$, (1.3.26) becomes the semi-discretized system (1.3.15) (or (1.3.20)). We will leave this as an exercise for the reader.

Note that, when restricted on U_h, the lumping operator P is surjective from U_h to V_h. Using this P, we rewrite Problem 1.3.1 as the following equivalent Galerkin finite element formulation.

Problem 1.3.2 *Find $u_h(t) \in U_h$ such that for all $v_h \in U_h$*

$$- (L_h(\dot{u}), L_h(v_h)) + B_h(u_h, v_h; t) = (L_h(f), L_h(v_h)), \qquad (1.3.28)$$

where $B_h(u_h, v_h; t) = A(u_h, L_h(v_h); t)$ with $A(\cdot, \cdot; t)$ defined in (1.3.27).

We first define functionals $\| \cdot \|_{0,h}$ and $\| \cdot \|_{1,h}$ on U_h by

$$\|v_h\|_{0,h}^2 := \sum_{j=1}^{N-1} v_j^2 l_j, \quad \|v_h\|_{1,h}^2 := \sum_{j=1}^{N-1} b_{j+1/2} S_{j+1/2} \frac{S_{j+1}^{\alpha_j} + S_j^{\alpha_j}}{S_{j+1}^{\alpha_j} - S_j^{\alpha_j}} (v_{j+1} - v_j)^2$$

$$(1.3.29)$$

for any $v_h = \sum_{j=0}^{N} v_j \phi_j \in U_h$, with $v_N = 0$. It is easy to show that $\|\cdot\|_{1,h}$ is a norm on U_h, because $(S_{j+1}^{\alpha_j} + S_j^{\alpha_j})/b_{j+1/2} > 0$ for any α_j $(= b_{j+1/2}/a)$. (The limiting case that $\alpha_j \to 0$ is given in (1.3.8)). Obviously, it is a weighted discrete energy norm on U_h. Using (1.3.29), we define the following weighted discrete H^1-norm on U_h $\|v_h\|_h^2 = \|v_h\|_{0,h}^2 + \|v_h\|_{1,h}^2$ with the convention that $v_0 = v_N = 0$. Thus, we have the following theorem.

Theorem 1.3.3 *Let Assumption 1.2.1 be fulfilled. If h is sufficiently small, then, for all $v_h \in U_h$, we have*

$$B_h(v_h, v_h; t) \geq C \|v_h\|_h^2 \qquad (1.3.30)$$

where C denotes a positive constant, independent of h and v_h.

Proof We omit the time variable t. Let $v_h = \sum_{j=1}^{N-1} v_i \phi_i \in U_h$. We have

$$B_h(v_h, v_h) = -\sum_{j=1}^{N-1} \left[x(ax \frac{\partial v_h}{\partial x} + \hat{b}v_h) \right]_{S_{j-1/2}}^{S_{j+1/2}} P(v_h) + (P(cv_h), P(v_h))$$

$$= -\sum_{j=1}^{N-1} \left(S_{j+\frac{1}{2}}(ax \frac{\partial v_h}{\partial x} + \hat{b}v_h)s_{j+\frac{1}{2}} - S_{j-\frac{1}{2}}(ax \frac{\partial v_h}{\partial x} + \hat{b}v_h)s_{j-\frac{1}{2}} \right) v_j + \sum_{j=1}^{N-1} c_j v_j^2 l_j.$$

Re-arranging the first sum and using (1.3.14) (with ρ_i given by (1.3.7)) and (1.3.12) (with $u_0 = 0$) we have

$$B_h(v_h, v_h) = \frac{S_{1/2}(a + b_{\frac{1}{2}})}{2} v_1^2 + \sum_{j=1}^{N-1} b_{j+\frac{1}{2}} S_{j+\frac{1}{2}} \frac{S_{j+1}^{\alpha_j} v_{j+1} - S_j^{\alpha_j} v_j}{S_{j+1}^{\alpha_j} - S_j^{\alpha_j}} (v_{j+1} - v_j)$$

$$+ \sum_{j=1}^{N-1} c_j v_j^2 l_j =: S_{1/2} \frac{(a + b_{1/2})}{2} v_1^2 + I + \sum_{j=1}^{N-1} c_j v_j^2 l_j, \qquad (1.3.31)$$

since $S_1 = 2 S_{1/2}$ and $v_0 = 0$. For the term I, we have

$$I = \sum_{j=1}^{N-1} b_{j+1/2} S_{j+\frac{1}{2}} \frac{S_{j+1}^{\alpha_j}(v_{j+1} - v_j) + (S_{j+1}^{\alpha_j} - S_j^{\alpha_j})v_j}{S_{j+1}^{\alpha_j} - S_j^{\alpha_j}} (v_{j+1} - v_j)$$

$$= \sum_{j=1}^{N-1} b_{j+1/2} S_{j+\frac{1}{2}} \frac{S_{j+1}^{\alpha_j}}{S_{j+1}^{\alpha_j} - S_j^{\alpha_j}} (v_{j+1} - v_j)^2 + \sum_{j=1}^{N-1} b_{j+\frac{1}{2}} S_{j+\frac{1}{2}} v_j (v_{j+1} - v_j)$$

$$= \sum_{j=1}^{N-1} b_{j+\frac{1}{2}} S_{j+\frac{1}{2}} \frac{S_{j+1}^{\alpha_j}}{S_{j+1}^{\alpha_j} - S_j^{\alpha_j}} (v_{j+1} - v_j)^2$$

$$+ \sum_{j=1}^{N-1} b_{j+\frac{1}{2}} S_{j+\frac{1}{2}} \left[-\frac{1}{2}(v_{j+1} - v_j)^2 + \frac{1}{2}(v_{j+1}^2 - v_j^2) \right]$$

$$= \frac{1}{2} \sum_{j=1}^{N-1} b_{j+\frac{1}{2}} S_{j+\frac{1}{2}} \frac{S_{j+1}^{\alpha_j} + S_j^{\alpha_j}}{S_{j+1}^{\alpha_j} - S_j^{\alpha_j}} (v_{j+1} - v_j)^2 + \frac{1}{2}[b_{1+\frac{1}{2}} S_{1+1/2}(v_2^2 - v_1^2)$$

$$+ b_{2+1/2} S_{2+\frac{1}{2}}(v_3^2 - v_2^2) + \cdots + b_{(N-1)+1/2} S_{(N-1)+\frac{1}{2}}(v_N^2 - v_{N-1}^2)]$$

$$= \frac{1}{2} \|v_h\|_{1,h}^2 + \frac{1}{2} \sum_{j=1}^{N-1} \left(S_{j-\frac{1}{2}} b_{j-\frac{1}{2}} - S_{j+\frac{1}{2}} b_{j+\frac{1}{2}} \right) v_j^2 - \frac{1}{2} S_{\frac{1}{2}} b_{\frac{1}{2}} v_1^2$$

$$= \frac{1}{2} \|v_h\|_{1,h}^2 - \frac{1}{2} \sum_{j=1}^{N-1} \left(S_{j-\frac{1}{2}} \frac{b_{j+\frac{1}{2}} - b_{j-\frac{1}{2}}}{l_j} + b_{j+\frac{1}{2}} \right) v_j^2 l_j - \frac{1}{2} S_{\frac{1}{2}} b_{\frac{1}{2}} v_1^2,$$

since $v_N = 0$. Note that r and σ in (1.2.13) are functions of t only. Therefore, substituting the above into (1.3.31) and using (1.3.29), (1.2.13) and (1.2.14) we have

$$B_h(v_h, v_h) = S_{1/2} \frac{(a + b_{\frac{1}{2}})}{2} v_1^2 + \frac{1}{2} \|v_h\|_{1,h}^2$$

$$+ \sum_{j=1}^{N-1} \left(c_j - \frac{S_{j-\frac{1}{2}}}{2} \frac{b_{j+\frac{1}{2}} - b_{j-\frac{1}{2}}}{l_j} - \frac{b_{j+\frac{1}{2}}}{2} \right) v_j^2 l_j - \frac{1}{2} S_{\frac{1}{2}} b_{\frac{1}{2}} v_1^2$$

$$= S_{1/2} \frac{a}{2} v_1^2 + \frac{1}{2} \|v_h\|_{1,h}^2$$

$$+ \sum_{j=1}^{N-1} \left(r + b_j - S_j d'(S_j) + \beta - \frac{S_{j-\frac{1}{2}}}{2} \frac{b_{j+\frac{1}{2}} - b_{j-\frac{1}{2}}}{l_j} - \frac{b_{j+\frac{1}{2}}}{2} \right) v_j^2 l_j$$

$$= S_{1/2} \frac{a}{2} v_1^2 + \frac{1}{2} \|v_h\|_{1,h}^2 + \sum_{j=1}^{N-1} \left[r + \frac{b_j}{2} + \frac{d_j - d_{j+\frac{1}{2}}}{2} - \frac{S_j}{2} d'(S_j) + \beta \right.$$

$$\left. - \frac{1}{2} \left(S_j d'(S_j) - S_{j-\frac{1}{2}} \frac{d_{j+\frac{1}{2}} - d_{j-\frac{1}{2}}}{l_j} \right) \right] v_j^2 l_j$$

$$\geq \frac{1}{2} \|v_h\|_{1,h}^2 + \frac{1}{2} \sum_{j=1}^{N-1} (3r - \sigma^2 - D'(S_j) + \beta) v_j^2 l_j$$

$$+ \sum_{j=1}^{N-1} \left[\frac{d_j - d_{j+\frac{1}{2}}}{2} - \frac{1}{2} \left(S_j d'(S_j) - S_{j-\frac{1}{2}} \frac{d_{j+\frac{1}{2}} - d_{j-\frac{1}{2}}}{l_j} \right) \right] v_j^2 l_j.$$

Note $d_j - d_{j+1/2}$ and $S_j d'(S_j) - S_{j-1/2}(d_{j+1/2} - d_{j-1/2})/l_j$ are of order $\mathcal{O}(h)$. When h is sufficiently small, the absolute values of these terms are smaller than, say, $\beta_0/2$. Thus, (1.3.30) follows from the above estimate and **Assumption** 1.2.1 because $(3r - D'(S_j) - \sigma^2 + \beta)/2 \geq 3r + \beta_0$. This completes the proof. $\qquad \square$

The lower bound of $B_h(v_h, v_h)$ is just a discrete analogue of that in Theorem 1.2.1. We remark that (1.3.30) implies that $B_h(\cdot, \cdot, t)$ is coercive with respect to $\|\cdot\|_h$. This result is crucial for the convergence analysis.

1.3.3.2　Full Discretization and Stability

Using the mesh for $(0, T)$ defined in Sect. 1.3.2, we pose the following problem:

Problem 1.3.3 *Find a sequence* $u_h^1, \ldots, u_h^K \in U_h$ *such that for* $k \in \{0, \ldots, K-1\}$

$$
\begin{cases}
\left(\dfrac{L_h(u_h^{k+1}) - L_h(u_h^k)}{-\Delta t_k}, L_h(v_h) \right) + B_h(\theta u_h^{k+1} + (1-\theta)u_h^k, v_h; t_{k+\theta}) \\
\qquad\qquad = (\theta L_h) f^{k+1}) + (1-\theta)L_h(f^k), v_h) \qquad \forall v_h \in U_h, \\
u_h^0 = g_3^I,
\end{cases}
$$

(1.3.32)

where $t_{k+\theta} := \theta t_{k+1} + (1-\theta)t_k = t_k + \theta \Delta t_k$, $f^k := f(t_k)$ *and* $g_3^I \in U_h$ *is the interpolant of the terminal condition* $g_3(S)$ *in* U_h.

It is easy to verify that if we choose $u_h^k = \sum_{j=1}^{N-1} u_j^k \phi(x)$ and $v_h = \phi_i$ for $i = 1, 2, \ldots, N-1$, in (1.3.32), (1.3.32) reduces to (1.3.21). We will leave this as an exercise for the reader.

The stability of the scheme is established in the following theorem:

Theorem 1.3.4 *If* $\theta \in [1/2, 1]$ *and* $g_3 \geq 0$ *in* $I \times (0, T)$, *then any solution to* (1.3.32) *satisfies*

$$
\|u_h^k\|_{0,h}^2 \leq \|g_3^I\|_{0,h}^2 + \frac{2T}{C_0} \sup_{s \in (0,T)} \|f(s)\|_{0,h}^2
$$

for a positive constant C_0, *independent of* i *and* u_h^i *for all* $i = 1, 2, \ldots, K$.

Proof In (1.3.32), we take the particular test function $v_h = u_h^\theta := \theta h_h^{k+1} + (1-\theta)u_h^k$. Since

$$
\theta u_h^{k+1} + (1-\theta)u_h^k = \left(\theta - \frac{1}{2} \right) \left[u_h^{k+1} - u_h^k \right] + \frac{1}{2} \left[u_h^{k+1} + u_h^k \right],
$$

we have, for $\theta \in [1/2, 1]$,

$$
\left(\frac{L_h(u_h^{k+1}) - u_h^k)}{-\Delta t_k}, L_h(u_h^\theta) \right) + B(u_h^\theta, u_h^\theta; t_{k+\theta})
$$

$$
= \frac{1}{-2\Delta t_k}(L_h(u_h^{k+1} - u_h^k), L_h(u_h^{k+1} + u_h^k)) + \frac{2\theta - 1}{-2\Delta t_k}(L_h(u_h^{k+1} - u_h^k), L_h(u_h^{k+1} - u_h^k))
$$

$$
+ B(u_h^\theta, u_h^\theta; t_{k+\theta}) \, ge \frac{1}{-2\Delta t_k} \left[\|u_h^{k+1}\|_{0,h}^2 - \|u_h^k\|_{0,h}^2 \right] + C_0 \|u_h^\theta\|_h^2,
$$

where C is the constant used in (1.3.30) and $\| \cdot \|_{0,h}$ is the norm defined in (1.3.29). Therefore we get from (1.3.32)

$$\frac{1}{-2\Delta t_k} \left[\|u_h^{k+1}\|_{0,h}^2 - \|u_h^k\|_{0,h}^2 \right] + C_0 \|u_h^\theta\|_h^2 \leq \left[\theta \|f^{k+1}\|_{0,h} + (1-\theta)\|f^k\|_{0,h} \right] \|u_h^\theta\|_h.$$

In the above we used the fact that $\|u_h^\Theta\|_{0,h} \leq \|u_h^\Theta\|_h$. Applying the ε-inequality $yz \leq \frac{y^2}{2\varepsilon} + \frac{\varepsilon z^2}{2}$ for any y, z and $\varepsilon > 0$ to the RHS of the above inequality with $\varepsilon = 2C_0$ we have

$$\frac{1}{-\Delta t_k} \left[\|u_h^{k+1}\|_{0,h}^2 - \|u_h^k\|_{0,h}^2 \right] \leq \frac{1}{C_0} \left[\|f^{k+1}\|_{0,h}^2 + \|f^k\|_{0,h}^2 \right],$$

or,

$$\|u_h^{k+1}\|_{0,h}^2 \leq \|u_h^k\|_{0,h}^2 + \frac{-\Delta t_k}{C_0} \left[\|f^{k+1}\|_{0,h}^2 + \|f^k\|_{0,h}^2 \right]$$

for $k = 0, 1, \ldots, K - 1$. Fro this recursive relationship we obtain

$$\|u_h^K\|_{0,h}^2 \leq \|u_h^0\|_{0,h}^2 + \frac{1}{C_0} \sum_{k=0}^{K-1} (-\Delta t_k) \left[\|f^{k+1}\|_{0,h}^2 + \|f^k\|_{0,h}^2 \right]$$

$$\leq \|u_h^0\|_{0,h}^2 + \frac{2T}{C_0} \sup_{s \in (0,T)} \|f(s)\|_{0,h}^2.$$

$\square$

1.3.3.3 Error Estimate

We now establish an upper bounds for the difference between the approximate and the exact solutions in $\| \cdot \|_h$. We start this discussion with the assumption that the spatial mesh is quasi-uniform as given below.

Assumption 1.3.1 There exists a constant $q_0 > 0$ such that $q_0^{-1} h_{i+1} \leq h_i \leq q_0 h_{i+1}$, $i = 0, \ldots, N - 1$.

Using this assumption we have the following lemma.

Lemma 1.3.1 *The discrete flux density ρ_h defined by (1.3.12), (1.3.14) and (1.3.7) can be written as*

$$\rho_h(v_h)|_{I_i} = \begin{cases} a(1+\alpha_0) S_{1/2} \dfrac{v_{h1} - v_{h0}}{h_0} + b_{1/2} v_{h0}, & i = 0, \\ a(1+\gamma_i) S_{i+1} \dfrac{v_{hi+1} - v_{hi}}{h_i} + b_{i+1/2} v_{hi}, & i = 1, \ldots, N-1, \end{cases}$$

where $\gamma_i := \alpha_i \frac{h_i S_{i+1}^{\alpha_i - 1}}{S_{i+1}^{\alpha_i} - S_i^{\alpha_i}} - 1$. *Furthermore, under Assumption 1.3.1, there exists a constant $C_\gamma > 0$ depending only on $\max_i \alpha_i$ and q_0 such that*

$$|\gamma_i x_{i+1}| \le C_\gamma h_i, \quad i = 1, \ldots, N - 1. \tag{1.3.33}$$

Proof Let us first consider the case $i = 0$. By (1.3.12),

$$\rho_h(v_h)|_{I_0} = \frac{1}{2} a h_0 \frac{v_{h1} - v_{h0}}{h_0} + \frac{1}{2} b_{1/2}[v_{h1} + v_{h0}].$$

Since $v_{h1} = v_{h0} + \dfrac{v_{h1} - v_{h0}}{h_0} h_0$, replacing v_{h1} in the last term of the above equation by this expression yields $\rho_h(v_h)|_{I_0} = (a + b_{1/2}) \frac{h_0}{2} \frac{v_{h1} - v_{h0}}{h_0} + b_{1/2} v_{h0}$. Finally it remains to observe that $a + b_{1/2} = a(1 + \alpha_0)$ and $h_0/2 = S_{1/2}$.

When $i = 1, \ldots, N - 1$, adding and subtracting $S_{i+1}^{\alpha_i} v_{hi}$ in the numerator of (1.3.7), we get

$$\rho_h(v_h)|_{I_i} = b_{i+1/2} \left[\frac{h_i S_{i+1}^{\alpha_i - 1}}{S_{i+1}^{\alpha_i} - S_i^{\alpha_i}} S_{i+1} \frac{v_{hi+1} - v_{hi}}{h_i} + v_{hi} \right].$$

But

$$b_{i+1/2} \frac{h_i S_{i+1}^{\alpha_i - 1}}{S_{i+1}^{\alpha_i} - S_i^{\alpha_i}} = a \alpha_i \frac{h_i S_{i+1}^{\alpha_i - 1}}{S_{i+1}^{\alpha_i} - S_i^{\alpha_i}} = a(1 + \gamma_i)$$

with γ_i defined in the Lemma.

Using a Taylor expansion we have

$$S_{i+1}^{\alpha_i} = S_i^{\alpha_i} + \alpha_i \xi^{\alpha_i - 1} h_i, \quad \text{where} \quad S_i < \xi < S_{i+1}, \tag{1.3.34}$$

from which we have $\gamma_i = \left(\frac{S_{i+1}}{\xi} \right)^{\alpha_i - 1} - 1 = \frac{S_{i+1}^{\alpha_i - 1} - \xi^{\alpha_i - 1}}{\xi^{\alpha_i - 1}}$.

Similarly, using the Taylor expansion $S_{i+1}^{\alpha_i - 1} = \xi^{\alpha_i - 1} + (\alpha_i - 1) \eta^{\alpha_i - 2}(S_{i+1} - \xi)$, where $\xi < \eta < S_{i+1}$, we have $\gamma_i = (\alpha_i - 1) \frac{(x_{i+1} - \xi)}{\xi} \left(\frac{\eta}{\xi} \right)^{\alpha_i - 2}$, and thus,

$$|\gamma_i x_{i+1}| \le |\alpha_i - 1| \frac{S_{i+1}}{\xi} \left(\frac{\eta}{\xi} \right)^{\alpha_i - 2} h_i \quad \text{with} \quad S_i < \xi < \eta < S_{i+1}.$$

To estimate the factor $(\eta/\xi)^{\alpha_i - 2}$, we have to consider the cases $\alpha_i < 2$ and $\alpha_i \ge 2$ separately. In the former case, we use that $\xi < \eta$ and so

$$\left(\frac{\eta}{\xi} \right)^{\alpha_i - 2} = \left(\frac{\xi}{\eta} \right)^{2 - \alpha_i} < \left(\frac{\eta}{\eta} \right)^{2 - \alpha_i} = 1.$$

In the latter case, we use $S_i < \xi$ and $\eta < S_{i+1}$ to obtain $\left(\frac{\eta}{\xi}\right)^{\alpha_i - 2} \leq \left(\frac{S_{i+1}}{S_i}\right)^{\alpha_i - 2}$.

Since $S_{i-1} \geq 0$ for all $i = 1, \ldots, N - 1$, we have the following estimate:

$$\frac{S_{i+1}}{S_i} = \frac{S_i + h_i}{S_i} = 1 + \frac{h_i}{S_i} = 1 + \frac{h_i}{S_{i-1} + h_{i-1}} \leq 1 + \frac{h_i}{h_{i-1}} \leq 1 + q_0$$

by Assumption 1.3.1.

Therefore $\left(\frac{\eta}{\xi}\right)^{\alpha_i - 2} \leq (1 + q_0)^{\alpha_i - 2}$. Similarly, $\frac{S_{i+1}}{\xi} < \frac{S_{i+1}}{S_i} \leq 1 + q_0$. So we finally obtain the estimate

$$|\gamma_i x_{i+1}| \leq \begin{cases} |\alpha_i - 1|(1 + q_0)h_i, & \alpha_i < 2, \\ |\alpha_i - 1|(1 + q_0)^{\alpha_i - 1}h_i, & \alpha_i \geq 2. \end{cases}$$

This is (1.3.33), and thus we have proved this lemma. $\qquad\square$

Now we are ready to prove the following consistency result.

Lemma 1.3.2 Let $w \in H^1_{0,w}(\Omega)$ be such that $\rho'(w, \cdot, t) \in L^2(\Omega)$ for all $t \in (0, T)$. Under Assumption (1.3.1), there exists a constant $C > 0$ depending only on q_0, C_γ and a and b, such that the following estimate holds:

$$|\rho_h(w_I, S_{i+1/2}, t) - \rho(w, S_{i+1/2}, t)| \leq C \int_{S_i}^{S_{i+1}} [|\rho'(w, S, t)| + |w'(S, t)| + |w(S, t)|] \, dS$$

for $i = 0, \ldots, N - 1$, where $w_I(\cdot, t)$ denotes the U_h-interpolant of $w(\cdot, t)$.

Proof We let $C > 0$ denote a generic positive constant, depending only on q_0, C_γ and on certain norms of the coefficients a and b.

By Lemma 1.3.1 and (1.3.2), for $i = 1, \ldots, N - 1$, we can write

$$\rho_h(w_I, S_{i+1/2}, \cdot) - \rho(w, S_{i+1/2}, \cdot)$$

$$= a(1 + \gamma_i) S_{i+1} \frac{w_{i+1} - w_i}{h_i} - a S_{i+1/2} w'(S_{i+1/2}) + b_{i+1/2}[w_i - w(S_{i+1/2})]$$

$$= a S_{i+1/2} \left[\frac{w_{i+1} - w_i}{h_i} - w'(S_{i+1/2})\right] + b_{i+1/2}[w_i - w(S_{i+1/2})]$$

$$+ a[S_{i+1} - S_{i+1/2} + \gamma_i S_{i+1}] \frac{w_{i+1} - w_i}{h_i} =: \vartheta_{1i} + \vartheta_{2i} + \vartheta_{3i}.$$

To estimate the first term, we use the following Taylor expansion with an integral remainder:

$$w(y) = w(S_{i+1/2}) + w'(S_{i+1/2})(y - S_{i+1/2}) + \int_{S_{i+1/2}}^{y} w''(x)(y - S) \, dS.$$

Replacing y by $y = S_{i+1}$ and S_i in the above expression, we have respectively

$$w_{i+1} = w(S_{i+1/2}) + w'(S_{i+1/2})\frac{h_i}{2} + \int_{S_{i+1/2}}^{S_{i+1}} w''(S)(S_{i+1} - S)dS'$$

$$w_i = w(S_{i+1/2}) - w'(S_{i+1/2})\frac{h_i}{2} - \int_{S_i}^{S_{i+1/2}} w''(S)(S_i - S)dS.$$

Therefore, from these equalities we have

$$\frac{w_{i+1} - w_i}{h_i} - w'(S_{i+1/2}) = \frac{1}{h_i}\int_{S_i}^{S_{i+1/2}} w''(S)(S_i - S)dS + \frac{1}{h_i}\int_{S_{i+1/2}}^{S_{i+1}} w''(S)(S_{i+1} - S)dS.$$

$$(1.3.35)$$

Note

$$\left|\int_{S_{i+1/2}}^{S_{i+1}} w''(S)(S_{i+1} - S)dS\right| \le \frac{h_i}{2}\int_{S_{i+1/2}}^{S_{i+1}} |w''|dS, \quad \left|\int_{S_i}^{S_{i+1/2}} w''(S)(S_i - S)dS\right| \le \frac{h_i}{2}\int_{S_i}^{S_{i+1/2}} |w''|dS.$$

Combining these estimates with (1.3.35) we obtain

$$\left|\frac{w_{i+1} - w_i}{h_i} - w'(x_{i+1/2})\right| \le \frac{1}{2}\int_{x_i}^{x_{i+1}} |w''|dx,$$

and so

$$|\vartheta_{1i}| \le \frac{a}{2}S_{i+1/2}\int_{S_i}^{S_{i+1}} |w''|dS = \frac{a}{2}\int_{S_i}^{S_{i+1}} \frac{S_{i+1/2}}{S}|Sw''|dS \le \frac{a}{2}\left(1 + \frac{q_0}{2}\right)\int_{S_i}^{S_{i+1}} |Sw''|dS,$$

where we used $\frac{S_{i+1/2}}{S} \le \frac{S_{i+1/2}}{S_i} \le 1 + \frac{h_i}{2S_i} \le 1 + \frac{q_0}{2}$ by Assumption (1.3.1). Since, by the definition of ρ, $axw'' = \rho' - (a+b)w' - b'w$, we obtain

$$|\vartheta_{1i}| \le \frac{1}{2}\left(1 + \frac{c}{2}\right)\int_{S_i}^{S_{i+1}} \left[|\rho'| + |a+b||w'| + |b'||w|\right]dS$$

$$\le C\int_{S_i}^{S_{i+1}} \left[|\rho'| + |w'| + |w|\right]dS.$$

To estimate ϑ_{2i}, we apply the formula $w(x_{i+1/2}) - w_i = \int_{S_i}^{S_{i+1/2}} w'dS$, and hence $|\vartheta_{2i}| \le |b_{i+1/2}|\int_{S_i}^{S_{i+1}} |w'|dS$. Similarly, since $w_{i+1} - w_i = \int_{S_i}^{S_{i+1}} w'dS$, we obtain from (1.3.33) in Lemma 1.3.1 that

$$|\vartheta_{3i}| \le a\left(\frac{h_i}{2} + C_\gamma h_i\right)\frac{|w_{i+1} - w_i|}{h_i} \le a\left(C_\gamma + \frac{1}{2}\right)\int_{S_i}^{S_{i+1}} |w'|dS.$$

Putting the above three estimates together, we have the following estimate:

$$|\rho_h(w_I, S_{i+1/2}, \cdot) - \rho(w, S_{i+1/2}, \cdot)| \le C\int_{S_i}^{S_{i+1}} \left[|\rho'| + |w'| + |w|\right]dS.$$

When $i = 0$, we have

$$\rho_h(w_I, S_{1/2}, \cdot) - \rho(w, S_{1/2}, \cdot)$$

$$= a(1 + \alpha_0) S_{1/2} \frac{w_1 - w_0}{h_0} - a S_{1/2} w'(S_{1/2}) + b_{1/2}[w_0 - w(S_{1/2})]$$

$$= \underbrace{a S_{1/2} \left[\frac{w_1 - w_0}{h_0} - w'(S_{1/2}) \right]}_{\vartheta_{10}} + \underbrace{b_{1/2}[w_0 - w(S_{1/2})]}_{\vartheta_{20}} + \underbrace{a \alpha_0 S_{1/2} \frac{w_1 - w_0}{h_0}}_{\vartheta_{30}}.$$

To estimate ϑ_{10}, we first proceed as in the general case and get

$$\frac{w_1 - w_0}{h_0} - w'(S_{1/2}) = \frac{1}{h_0} \int_{S_0}^{S_{1/2}} w''(S)(S_0 - S)dS + \frac{1}{h_0} \int_{S_{1/2}}^{S_1} w''(S)(S_1 - S)dS.$$

Therefore

$$\left| \frac{w_1 - w_0}{h_0} - w'(S_{1/2}) \right| \leq \frac{1}{h_0} \int_{S_0}^{S_{1/2}} |Sw''(S)|dS + \frac{1}{h_0} \int_{S_{1/2}}^{S_1} |w''(S)||S_1 - S|dS.$$

For $S \in [S_{1/2}, S_1]$, it is easy to see that

$$|S_1 - S| = S_1 - S = S \left(\frac{S_1}{S} - 1 \right) \leq S \left(\frac{S_1}{S_{1/2}} - 1 \right) = S,$$

and we obtain

$$\left| \frac{w_1 - w_0}{h_0} - w'(S_{1/2}) \right| \leq \frac{1}{h_0} \int_{S_0}^{S_1} |Sw''|dS.$$

It follows that

$$|\vartheta_{10}| \leq a \frac{S_{1/2}}{h_0} \int_{S_0}^{S_1} |Sw''|dS = \frac{1}{2} \int_{S_0}^{S_1} |aSw''(S)|dS \leq C \int_{S_0}^{S_1} \left[|\rho'| + |w'| + |w| \right] dS.$$

The term ϑ_{20} can be estimated as in the general case, i.e. $|\vartheta_{20}| \leq |b_{1/2}| \int_{S_0}^{S_1} |w'|dS$. Finally, we have

$$|\vartheta_{30}| = a|\alpha_0| S_{1/2} \frac{|w_1 - w_0|}{h_0} = \frac{|b_{1/2}|}{2} \int_{S_0}^{S_1} |w'|dS.$$

In summary, we get

$$|\rho_h(w_I, S_{1/2}, \cdot) - \rho(w, S_{1/2}, \cdot)| \leq C \int_{x_0}^{x_1} \left[|\rho'| + |w'| + |w| \right] dS. \qquad \square$$

We are ready to prove the main convergence result for the FVM in the previous subsections. For simplicity and clarity, we only consider the case that $\theta = 1$, i.e., the fully implicit or backward Euler's scheme. We also assume that the solution u to Problem 1.2.1 is sufficiently smooth to avoid intensive discussion on the regularity requirements. We will estimate $R^k := u_I(t_k) - u_h^k$ in a discrete energy norm to be defined, where $u_I(\cdot, t)$ is the U_h-interpolation of $u(\cdot, t)$ defined above. In what follows, we use $C > 0$ to denote a generic positive constant, independent of both h_i and Δt_k for all feasible i and k.

For any $v \in C(\bar{I})$, multiplying (1.2.12) by $L_h(v)$ and integrating the 2nd term by parts, we have

$$(-\dot{u}, L_h(v)) + \hat{B}_h(u, v); t) = (f, L_h(v)), \tag{1.3.36}$$

where L_h is the operator defined in (1.3.25) and

$$\hat{B}_h(u, v; t) = -\sum_{i=1}^{N-1} [S\rho(u, x, t)]_{S_{i-1/2}}^{S_{i+1/2}} v(S_i) + (cu, L_h(v)).$$

Adding and subtracting appropriate terms and using (1.3.36) with the test functions $v = v_h \in U_h$ and (1.3.32), we easily derive the following equation with respect to R^{k+1}:

$$\left(\frac{L_h(R^{k+1}) - L_h(R^k)}{-\Delta t_k}, L_h(v_h) \right) + B_h(R^{k+1}, v_h; t_{k+1})$$

$$= \left(\frac{L_h(u_I(t_{k+1})) - L_h(u_I(t_k))}{-\Delta t_k}, v_h \right) + B_h(u_I(t_{k+1}), v_h; t_{k+1})$$

$$- \left(\frac{L_h(u_h^{k+1}) - L_h(u_h^k)}{-\Delta t_k}, v_h \right) - B_h(u_h^{k+1}, v_h; t_{k+1})$$

$$= -(\dot{u}(t_{k+1}), L_h(v_h)) + (\dot{u}(t_{k+1}), L_h(v_h)) + \left(\frac{L_h(u_I(t_{k+1})) - L_h(u(t_k))}{-\Delta t_k}, v_h \right)$$

$$+ \hat{B}_h(u(t_{k+1}), v_h; t_{m+1}) - \hat{B}_h(u(t_{k+1}), v_h; t_{m+1}) + B_h(u_I(t_{k+1}), v_h; t_{k+1})$$

$$- \left(\frac{L_h(u_h^{k+1}) - L_h(u_h^k)}{-\Delta t_k}, L_h(v_h) \right) - B_h(u_h^{k+1}, v_h; t_{k+1})$$

$$= \underbrace{\left[\left(\frac{L_h(u_I(t_{k+1})) - L_h(u_I(t_k))}{-\Delta t_k}, v_h \right) + (\dot{u}(t_{k+1}), L_h(v_h)) \right]}_{Y_1^k}$$

$$+ \underbrace{\left[B_h(u_I(t_{k+1}), v_h; t_{k+1}) - \hat{B}_h(u(t_{k+1}), v_h; t_{k+1}) \right]}_{Y_2^k}$$

$$+ \underbrace{\left[(f^{k+1} L_h(v_h)) - (L_h(f^{k+1}), L_h(v_h)) \right]}_{Y_3^k} =: Y_1^k + Y_2^k + Y_3^k. \tag{1.3.37}$$

Now we estimate the terms $|Y_j^k|$, $j = 1, 2, 3$ separately.

(i) Estimation of $|Y_1^k|$:

Let $w^k := \frac{L_h(u(t_{k+1})) - L_h(u(t_k))}{-\Delta t_k} + \dot{u}(t_{k+1})k$. Then, by Cauchy–Schwarz inequality, we have $|Y_1^k| \le \|w^k\|_0 \|v_h\|_{0,h}$. Thus, we need to derive a bound for $\|w^k\|_0$. A simple algebraic manipulation yields

$$w^k = \frac{1}{-\Delta t_k}\left[(L_h(u(t_{k+1})) - u(t_{k+1})) - (L_h(u(t_k)) - u(t_k))\right]$$
$$+ \frac{1}{-\Delta t_k}(u(t_{k+1}) - u(t_k)) + \dot{u}(t_{k+1}). \tag{1.3.38}$$

To estimate the first term on the RHS of (1.3.38), we use a Taylor expansion with integral remainder for it to obtain

$$\frac{1}{-\Delta t_k}\left[(L_h(u(t_{k+1})) - u(t_{k+1}) - (L_h(u(t_k)) - u(t_k)\right] = \frac{1}{-\Delta t_k}\int_{t_k}^{t_{k+1}}\frac{d}{ds}[(L_h(u(s)) - u(s)]ds,$$

and so

$$\left\|\frac{1}{-\Delta t_k}\left[(L_h(u(t_{k+1})) - u(t_{k+1}) - (L_h(u(t_k)) - u(t_k)\right]\right\|_0 \le \frac{1}{-\Delta t_k}\int_{t_{k+1}}^{t_k}\|L_h(\dot{u}(s)) - \dot{u}(s)\|_0 ds.$$

For the the second term on the RHS of (1.3.38), we have, using a Taylor expansion with an integral remainder,

$$u(t_k) = u(t_{k+1}) + \dot{u}(t_{k+1})(-\Delta t_k) + \int_{t_{k+1}}^{t_k}(t_k - s)\ddot{u}(s)ds,$$

from which, we get

$$\frac{1}{-\Delta t_k}(u(t_{k+1}) - u(t_k)) + \dot{u}(t_{k+1}) = \frac{1}{-\Delta t_k}\int_{t_k}^{t_{k+1}}(t_k - s)\ddot{u}(s)ds.$$

Taking the norm on both sides, we have the following estimate for the second term on the RHS of (1.3.38):

$$\left\|\frac{1}{-\Delta t_k}(u(t_{k+1}) - u(t_k)) + \dot{u}(t_{k+1})\right\|_0 \le \int_{t_{k+1}}^{t_k}\|\ddot{u}(s)\|_0 ds.$$

Combining the estimates for the two terms on the RHS of (1.3.38) and using the triangular inequality, we obtain the following estimate for $|Y_1^k|$:

$$|Y_1^m| \le \|w^m\|_0 \|v_h\|_{0,h} \le Q_1^m(\Delta t_k, h)\|v_h\|_{0,h}, \tag{1.3.39}$$

where

$$Q_1^k(\Delta t_k, h) := \frac{1}{-\Delta t_k} \int_{t_{k+1}}^{t_k} \|L_h(\dot{u}(s)) - \dot{u}(s)\|_0 ds + \int_{t_{k+1}}^{t_k} \|\ddot{u}(s)\|_0 ds. \quad (1.3.40)$$

(ii) Estimation of $|Y_2^k|$:

From the definitions of B_h and $\hat{B}_h$ in (1.3.28) and (1.3.36) respectively we have

$$Y_2^k = B_h(u_I(t_{k+1}), v_h; t_{k+1}) - \hat{B}_h(u(t_{k+1}), v_h; t_{k+1})$$

$$= -\sum_{i=1}^{N-1} \left[S[\rho_h(u_I, S, t_{k+1}) - \rho(u, S, t_{k+1})] \right]_{S_{i-1/2}}^{S_{i+1/2}} v_{hi} + (L_h(cu_I) - cu, L_h(v_h)),$$

where $v_{hi} = v_h(S_i)$. Re-arranging the first sum and using the boundary conditions $v_{h0} = v_{hN} = 0$, we get

$$|Y_2^k| \le \left| \sum_{i=0}^{N-1} S_{i+1/2}[\rho_h(u_I, S_{i+1/2}, t_{k+1}) - \rho(u, S_{i+1/2}, t_{k+1})](v_{hi} - v_{hi+1}) \right|$$

$$+ |(L_h(cu_I) - cu, L_h(v_h))| =: \delta_1 + \delta_2.$$

For δ_1, we have

$$\delta_1 \le |\rho_h(u_I, S_{1/2}, t_{k+1}) - \rho(u, S_{1/2}, t_{k+1})|S_{1/2}|v_{h0} - v_{h1}|$$

$$+ \sum_{i=1}^{N-1} |\rho_h(u_I, S_{i+1/2}, t_{k+1}) - \rho(u, S_{i+1/2}, s)|S_{i+1/2}|v_{hi} - v_{hi+1}|. \quad (1.3.41)$$

Since $S_{1/2} = h_0/2$ and $v_{h0} = 0$, we can write

$$|\rho_h(u_I, S_{1/2}, t_{k+1}) - \rho(u, S_{1/2}, t_{k+1})|S_{1/2}|v_{h0} - v_{h1}|$$

$$\le Ch_0 \left\{ \int_{S_0}^{S_1} [|\rho'| + |w'| + |w|] dS \right\} |v_{h1}| \le Ch_0 \left\{ \int_{S_0}^{S_1} [|\rho'| + |w'| + |w|]^2 dS \right\}^{1/2} \sqrt{h_0 v_{h1}^2}. \quad (1.3.42)$$

The estimate of the remaining sum in (1.3.41) is more completed. We start with the consideration of the terms $S_{i+1/2}|v_{hi} - v_{hi+1}|$ for $i = 1, \ldots N - 1$. By a simple algebraic manipulation, it holds

$$S_{i+1/2}|v_{hi} - v_{hi+1}| = \left(\frac{S_{i+1/2}}{b_{i+1/2}} \right)^{1/2} \left(\frac{S_{i+1}^{\alpha_i} - S_i^{\alpha_i}}{S_{i+1}^{\alpha_i} + S_i^{\alpha_i}} \right)^{1/2} \times$$

$$\times \left(S_{i+1/2} b_{i+1/2} \frac{S_{i+1}^{\alpha_i} + S_i^{\alpha_i}}{S_{i+1}^{\alpha_i} - S_i^{\alpha_i}} \right)^{1/2} |v_{hi} - v_{hi+1}|. \quad (1.3.43)$$

Now, using the Taylor expansion (1.3.34), we have $S_{i+1}^{\alpha_i} - S_i^{\alpha_i} = \alpha_i \xi^{\alpha_i-1} h_i$, $\xi \in (S_i, S_{i+1})$. Hence

$$\frac{S_{i+1}^{\alpha_i} - S_i^{\alpha_i}}{S_{i+1}^{\alpha_i} + S_i^{\alpha_i}} = \frac{\alpha_i h_i}{\xi} \frac{\xi^{\alpha_i}}{S_{i+1}^{\alpha_i} + S_i^{\alpha_i}}. \tag{1.3.44}$$

If $\alpha_i < 0$, then $\xi^{\alpha_i} < S_i^{\alpha_i}$, and so $\frac{\xi^{\alpha_i}}{S_{i+1}^{\alpha_i}+S_i^{\alpha_i}} \le \frac{S_i^{\alpha_i}}{S_{i+1}^{\alpha_i}+S_i^{\alpha_i}} \le 1$.

Analogously, if $\alpha_i \ge 0$, then $\xi^{\alpha_i} < S_{i+1}^{\alpha_i}$. Thus, $\frac{\xi^{\alpha_i}}{S_{i+1}^{\alpha_i}+S_i^{\alpha_i}} \le \frac{S_{i+1}^{\alpha_i}}{S_{i+1}^{\alpha_i}+S_i^{\alpha_i}} \le 1$.
Combining these estimates with (1.3.44), we have

$$\left| \frac{S_{i+1}^{\alpha_i} - S_i^{\alpha_i}}{S_{i+1}^{\alpha_i} + S_i^{\alpha_i}} \right| \le \frac{|\alpha_i| h_i}{\xi} = \frac{S_{i+1/2}}{\xi} \frac{|\alpha_i| h_i}{S_{i+1/2}} \le \left(1 + \frac{q_0}{2}\right) \frac{|\alpha_i| h_i}{S_{i+1/2}}.$$

So, the 1st product on the RHS of (1.3.43) can be estimated as follows:

$$\frac{S_{i+1/2}}{|b_{i+1/2}|} \left| \frac{S_{i+1}^{\alpha_i} - S_i^{\alpha_i}}{S_{i+1}^{\alpha_i} + S_i^{\alpha_i}} \right| \le \left(1 + \frac{c}{2}\right) \frac{h_i}{a} \le C h_i,$$

where $C > 0$ denotes a generic positive constant, independent of h_i's and Δt_k's. Using this estimate, we have from (1.3.43)

$$\sum_{i=1}^{N-1} |\rho_h(u_I, S_{i+1/2}, t_{k+1}) - \rho(u, S_{i+1/2}, t_{k+1})| S_{i+1/2} |v_{hi} - v_{hi+1}|$$

$$\le C \sum_{i=1}^{N-1} \sqrt{h_i} \left\{ \int_{S_i}^{S_{i+1}} [|\rho'| + |w'| + |w|] \, dS \right\} \left(S_{i+1/2} b_{i+1/2} \frac{S_{i+1}^{\alpha_i} + S_i^{\alpha_i}}{S_{i+1}^{\alpha_i} - S_i^{\alpha_i}} \right)^{1/2} |v_{hi} - v_{hi+1}|$$

$$\le C \sum_{i=1}^{N-1} h_i \left\{ \int_{S_i}^{S_{i+1}} [|\rho'| + |w'| + |w|]^2 \, dS \right\}^{1/2} \left(S_{i+1/2} b_{i+1/2} \frac{S_{i+1}^{\alpha_i} + S_i^{\alpha_i}}{S_{i+1}^{\alpha_i} - S_i^{\alpha_i}} \right)^{1/2} |v_{hi} - v_{hi+1}|$$

$$\le C h \sum_{i=1}^{N-1} \left\{ \int_{S_i}^{S_{i+1}} [|\rho'| + |w'| + |w|]^2 \, dS \right\}^{1/2} \left(S_{i+1/2} b_{i+1/2} \frac{S_{i+1}^{\alpha_i} + S_i^{\alpha_i}}{S_{i+1}^{\alpha_i} - S_i^{\alpha_i}} \right)^{1/2} |v_{hi} - v_{hi+1}|$$

$$\le C h \left\{ \sum_{i=1}^{N-1} \int_{S_i}^{x_{i+1}} [|\rho'| + |w'| + |w|]^2 \, dS \right\}^{1/2} \left\{ \sum_{i=1}^{N-1} S_{i+1/2} b_{i+1/2} \frac{S_{i+1}^{\alpha_i} + S_i^{\alpha_i}}{S_{i+1}^{\alpha_i} - S_i^{\alpha_i}} |v_{hi} - v_{hi+1}|^2 \right\}^{1/2}$$

$$= C h \left\{ \int_{S_1}^{S_N} [|\rho'| + |w'| + |w|]^2 \, dS \right\}^{1/2} \|v_h\|_{1,h}.$$

Together with the estimate (1.3.42) we have (recalling Assumption (1.3.1))

$$\delta_1 \le C h_0 \left\{ \int_{S_0}^{S_1} [|\rho'| + |w'| + |w|]^2 \, dS \right\}^{1/2} \sqrt{h_0 v_{h1}^2}$$

$$+ C h \left\{ \int_{S_1}^{S_N} [|\rho'| + |w'| + |w|]^2 \, dS \right\}^{1/2} \|v_h\|_{1,h}$$

$$\le C h \left\{ \int_{S_0}^{S_N} [|\rho'| + |w'| + |w|]^2 \, dS \right\}^{1/2} \left\{ l_1 v_{h1}^2 + \|v_h\|_{1,h}^2 \right\}^{1/2}$$

$$\le C h (|\rho|_1 + \|w\|_1) \|v_h\|_h,$$

where $|w|_1 = \|w'\|_0$ denotes the 1st-order semi-norm of w. For the second term δ_2, we have by a standard argument $\delta_2 \leq Ch \left(|c(s)|_1 + |w|_1\right) \|v_h\|_{0,h}$. Combining the bounds on δ_1 and δ_2 yields $|Y_2^k| \leq Ch \left(|c(s)|_1 + |\rho|_1 + \|w\|_1\right) \|v_h\|_h$.

(iii) Estimation of $|Y_3^k|$:

For Y_3^k, we simply have

$$|Y_3^k| = |(f^{k+1} - L_h(f^{k+1}), L_h(v_h))| = \sum_{i=1}^{N-1} v_{hi} \int_{S_{i-1/2}}^{S_{i+1/2}} \left(f(S, t_{k+1}) - f_i^{k+1}\right) dS$$

$$\leq Ch|f|_1 \sum_{i=1}^{N-1} v_{hi} l_i \leq Ch|f|_1 \|v_h\|_{0,h}. \tag{1.3.45}$$

Combining the estimates (1.3.39)–(1.3.45), we get from (1.3.37)

$$\left(\frac{L_h(R^{k+1}) - L_h(R^k)}{-\Delta t_k}, L_h(v_h)\right) + B_h(R^{k+1}, v_h; t_{k+1})$$

$$\leq C\left[(Q_1^k + h|f|_1) \|v_h\|_{0,h} + h \left(|c|_1 + |\rho|_1 + \|u\|_1\right) \|v_j\|_h\right] \leq Q^k(\Delta t_k, h) \|v_h\|_h, \tag{1.3.46}$$

because $\|v\|_{0,h} \leq \|v_h\|_h$, where Q_1^k is defined in (1.3.40) and

$$Q^k(\Delta t_k, h) := C \left(Q_1^k(\Delta t_k, h) + (|c|_1 + |\rho|_1 + \|u\|_1 + |f|_1)h\right). \tag{1.3.47}$$

We now choose in (1.3.46) the particular test function: $v_h = R^{k+1}$. Applying the same argument as in the proof of Theorem 1.3.4 (with $\theta = 1$), we get the following estimate:

$$\left(\frac{L_h(R^{k+1}) - L_h(R^k)}{-\Delta t_k}, L_h(R^{k+1})\right) + B_h(R^{k+1}, R^{k+1}; t_{k+1})$$

$$\geq \frac{1}{-2\Delta t_k} \left(\|R^{k+1}\|_{0,h}^2 - \|R^k\|_{0,h}^2\right) + C_0 \, \|R^{k+1}\|_h^2.$$

Combining this with (1.3.46) leads to

$$\frac{1}{-2\Delta t_k} \left(\|R^{k+1}\|_{0,h}^2 - \|R^k\|_{0,h}^2\right) + C_0 \|R^{k+1}\|_h^2 \leq Q^k(\Delta t_k, h) \|v_h\|_h.$$

Applying the ε-inequality given above to the RHS of the above with $\varepsilon = C_0$, we get

$$\frac{1}{-2\Delta t_k} \left(\|R^{k+1}\|_{0,h}^2 - \|R^k\|_{0,h}^2\right) + C_0 \|R^{k+1}\|_h^2 \leq \frac{1}{2C_0}[Q^k(\Delta t_k, h)]^2 + \frac{C_0}{2} \|R^{k+1}\|_h^2,$$

and thus it follows that

$$\frac{1}{-2\Delta t_k}\left(\|R^{k+1}\|_{0,h}^2 - \|R^k\|_{0,h}^2\right) + \frac{C_0}{2}\|R^{k+1}\|_h^2 \le \frac{1}{2C_0}[Q^k(\Delta t_k,h)]^2.$$

Multiplying this inequality by $-2\Delta t_k(>0)$, we obtain

$$\|R^{k+1}\|_{0,h}^2 - \|R^k\|_{0,h}^2 - C_0\Delta t_k\|R^{k+1}\|_h^2 \le \frac{-\Delta t_k}{C_0}[Q^k(\Delta t_k,h)]^2.$$

Summing up both sides of the above from $k=0$ to $k=K-1$ and using the definition of Q^k in (1.3.47) we have

$$\|R^K\|_{0,h}^2 + C_0\sum_{k=0}^{K-1}(-\Delta t_k)\|R^{k+1}\|_h^2 \le \frac{1}{C_0}\sum_{k=0}^{K-1}(-\Delta t_k)[Q^k(\Delta t_k,h)]^2$$

$$\le C\left(\sum_{k=0}^{K-1}(-\Delta t_k)[Q_1^k(\Delta t_k,h)]^2 + (|c|_1 + |\rho|_1 + \|u\|_1 + |f|_1)^2 h^2 T\right). \quad (1.3.48)$$

It remains to estimate the sum on the RHS of (1.3.48).

From (1.3.40) we get

$$[Q_1^k(\Delta t_k,h)]^2$$

$$\le 2\left[\frac{1}{(\Delta t_k)^2}\left\{\int_{t_{k+1}}^{t_k}\|L_h(\dot{u}(s)) - \dot{u}(s)\|_0\,ds\right\}^2 + \left\{\int_{t_{k+1}}^{t_k}\|\ddot{u}(s)\|_0\,ds\right\}^2\right]$$

$$\le 2\left[\frac{1}{-\Delta t_k}\int_{t_{k+1}}^{t_k}\|(L_h - I)\dot{u}(s)\|_0^2\,ds - \Delta t_k\int_{t_{k+1}}^{t_k}\|\ddot{u}(s)\|_0^2\,ds\right],$$

and, therefore,

$$\sum_{k=0}^{K-1}(-\Delta t_k)[Q_1^k(\Delta t_k,h)]^2 \le 2\sum_{k=0}^{K-1}\left[\int_{t_{k+1}}^{t_k}\|L_h(\dot{u}(s)) - \dot{u}(s)\|_0^2\,ds + (\Delta t_k)^2\int_{t_{k+1}}^{t_k}\|\ddot{u}(s)\|_0^2\,ds\right]$$

$$\le 2\left[\int_0^T\|L_h(\dot{u}(s)) - \dot{u}(s)\|_0^2\,ds + (\Delta t)^2\int_0^T\|\ddot{u}(s)\|_0^2\,ds\right],$$

where $\Delta t := \max_{k=0,\dots,K-1}|\Delta t_k|$. By a standard argument, $\|L_h(\dot{u}(s)) - \dot{u}(s)\|_0 \le h|\dot{u}(s)|_1$, and so we arrive at

$$\sum_{k=0}^{K-1}(-\Delta t_k)[Q_1^k(\Delta t_k,h)]^2 \le 2\left[h^2\|\dot{u}\|_{L^2(0,T;H^1(I))}^2 + (\Delta t)^2\|\ddot{u}\|_{L^2(0,T;L^2(I))}^2\right].$$

$$(1.3.49)$$

Replacing the sum in (1.3.48) by the upper bound in (1.3.49), we finally get the estimate

$$\|R^K\|_{0,h}^2 + C_0 \sum_{k=0}^{K-1} (-\Delta t_k) \|R^{k+1}\|_h^2 \le C(h^2 + \Delta t^2). \tag{1.3.50}$$

To summarize, we have proven, in the above, the following theorem:

Theorem 1.3.5 *Let u and $\{u_h^k \in U_h : k = 1, 2, \ldots, K\}$ be respectively the solutions to Problems 1.3.2 and 1.3.3 with $\theta = 1$ and assume (1.3.1) holds. If u is such that is sufficiently smooth, then there exists a positive constant $C > 0$, independent of h and Δt, such that*

$$\|u_I(t_0) - u_h^K\|_{0,h} + \left(\sum_{k=0}^{K-1} |\Delta t_k| \|u_I(t_k) - u_h^k\|_h^2 \right)^{1/2} \le C(h + \Delta t), \tag{1.3.51}$$

where $u_I(t_k)$ denotes the U_h-interpolant of $u(\cdot, t_k)$ for $k = 0, 1, \ldots, K$.

Proof The estimate (1.3.51) follows immediately from (1.3.50). □

Remark 1.3.3 Since $\| \cdot \|_{0,h}$ in (1.3.51) depends only on the nodal values of a given function, Theorem 1.3.5 still holds if we replace u_I by u, as both of them have the same nodal values. We also comment, in Theorem 1.3.5 we used a 'vague' assumption that 'u is sufficiently smooth' to avoid introducing more spaces and norms.

1.4 Numerical Experiments

We now present some numerical examples to demonstrate the usefulness of the finite volume method developed in the previous sections. For all the examples given below, we choose $S_{\max} = 100$ and $T = 1$. The uniform mesh with 51×51 mesh nodes is used for solving all the tests below.

Example 1.1 European call option with $K = 50$ and market parameters $\sigma = 0.4$, $r = 0.03$, $d = 0.02$. The payoff and boundary conditions are given in (1.2.6) and (1.2.9)–(1.2.10) respectively.

The computed option value and its derivative with respect to S (i.e., Δ) are plotted in Fig. 1.2.

Example 1.2 European call option with $K = 50$. The market parameters are chosen to be $\sigma = 0.4 + 0.2 \sin(10t)$, $r = 0.06 + 0.02 \sin(10t)$ and $d = 0.0005S$, and the payoff and boundary conditions are given in (1.2.6)–(1.2.10).

The computed option value V and the corresponding Δ are plotted in Fig. 1.3.

Example 1.3 European call option with the market parameters $\sigma = 0.4$, $r = 0.04$, $d = 0.02S$ and strike price $K = 50$. This is a Cash-Or-Nothing option and its payoff condition is given in (1.2.7) with $B = 1$. The boundary conditions are $V(0, t) = 0$ and $V(S_{\max}, t) := \exp(-r(T - t))$ for $t \in [0, T)$.

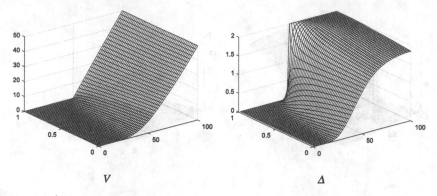

Fig. 1.2 Computed value V and Δ for Example 1.1

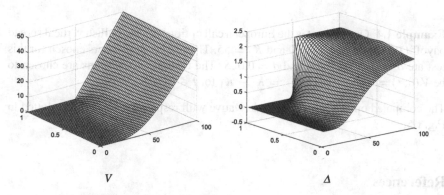

Fig. 1.3 Computed value V and Δ for Example 1.2

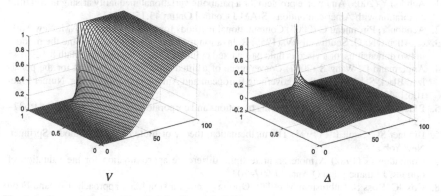

Fig. 1.4 Computed value V and Δ for Example 1.3

The computed option value V and Δ are plotted in Fig. 1.4. As can be seen from Fig. 1.4, the derivative of V does not exist at $S = 50$.

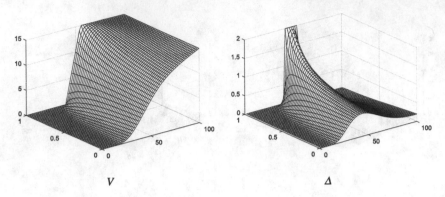

$$V \qquad\qquad\qquad\qquad \Delta$$

Fig. 1.5 Computed value V and Δ for Example 1.4

Example 1.4 Our last test is the European call option with the bullish vertical spread payoff (1.2.8) where $K_1 = 40$ and $K_2 = 55$. The market parameters chosen for this test are $\sigma = 0.4$, $r = 0.04$ and $d = 0.02S$. The boundary conditions are chosen to be $V(0, t) = 0$ and $V(S_{\max}, t) := K_2 - K_1$ for $t \in [0, T)$.

The computed option value and its derivative with respect to S and Δ are plotted in Fig. 1.5.

References

1. Achdou Y (2005) An inverse problem for a parabolic variational inequality arising in volatility calibration with American options. SIAM J Control Optim 43:1583–1615
2. Achdou Y, Pironneau O (2005) Computational methods for option pricing. SIAM, New York
3. de Allen DN, G, Southwell RV, (1955) Relaxation methods applied to determine the motion, in two dimensions, of a viscous fluid past a fixed cylinder. Quart J Mech Appl Math 8:129–145
4. Angermann L, Wang S (2007) Convergence of a fitted finite volume method for the penalized Black-Scholes equations governing European and American option pricing. Numer Math 106:1–40
5. Black F, Scholes M (1973) The pricing of options and corporate liabilities. J Polit Econ 81:637–659
6. Brenner SC, Scott LR (1994) The mathematical theory of finite element methods. Springer, New Yok
7. Courtadon G (1982) A more accurate finite difference approximation for the valuation of options. J Financ Econ Q Anal 17:697–703
8. Cox JC, Ross S, Rubinstein M (1979) Option pricing: a simplified approach. J Financ Econ 7:229–264
9. Dewynne JN, Howison SD, Rupf I, Wilmott P (1993) Some mathematical results in the pricing of American options. Euro J Appl Math 4:381–398
10. Duffy D (2006) Finite difference methods in financial engineering - a partial differential equation approach. Wiley, New Work
11. Itô K (1951) On stochastic differential equations. Mem Amer Math Soc 4:1–51
12. Haslinger J, Miettinen M (1999) Finite Element Method for Hemivariational Inequalities. Kluwer Academic Publisher, Dordrecht

13. Hull J, White A (1996) Hull-White on derivatives. Risk Publications, London
14. Kufner A (1985) Weighted Sobolev spaces. Wiley, New York
15. Miller JJH, Wang S (1994) A new non-conforming Petrov-Galerkin method with triangular elements for a singularly perturbed advection-diffusion problem. IMA J Numer Anal 14:257–27
16. Miller JJH, Wang S (1994) An exponentially fitted finite element volume method for the numerical solution of 2D unsteady incompressible flow problems. J Comput Phys 115:56–64
17. Renardy M, Rogers RC (2004) An introduction to partial differential equations. Texts in applied mathematics, 2nd edn., vol. 13. Springer, New York
18. Rogers LCG, Tallay D (1997) Numerical methods in finance. Cambridge University Press, Cambridge
19. Varga RS (1962) Matrix Iterative Analysis. Prentice-Hall, Englewood Cliffs
20. Vazquez C (1998) An upwind numerical approach for an American and European option pricing model. Appl Math & Comput 97:273–286
21. Wang S (2004) A novel fitted finite Volume method for the Black-Scholes equation governing option pricing. IMA J Numer Anal 24:699–720
22. Wilmott P, Dewynne J, Howison J (1993) Option pricing: mathematical models and computation. Oxford Financial Press, Oxford

Chapter 2
American Options on One Asset

Abstract In this chapter, we first give a brief account of the derivation of the differential Linear Complementarity Problem (LCP) governing American put option valuation. We then write this LCP as a variational inequality, which is shown to be uniquely solvable. A partial differential equation with a nonlinear penalty term is proposed to approximate the LCP. We prove that the penalty equation is uniquely solvable and its solution converges to the weak solution to the LCP exponentially. The finite volume method in Chap. 1 is used for the penalty equation. A Newton's algorithm is proposed to solve the discretized nonlinear system. Numerical experimental results are presented to demonstrate this method produces financially meaningful numerical solutions to the American put option pricing problem.

Keywords American option valuation · Linear complementarity problem ·
Variational inequality · Power penalty method · Convergence.

2.1 The Differential LCP and Its Solvability

An American option is one which can be exercised any time prior to or on the maturity/expiry date T. Therefore, a question that arises is what is the optimal time to exercise an American option? A solution to an American option pricing problem contains two components—the the option value and optimal exercise curve which divides the solution domain in space and time into two sub-domains, so that in one sub-domain, the option should be exercised and in the other, it should be held.

An American call option has the same value as its European counterpart. An intuitive explanation for this is that, at time $t < T$, the option should be held if its holder thinks the underlying stock price will increase. On the other hand, if the option holder thinks the stock price will decrease, he/she can short the stock without exercising the option, and the risk of this shorting is covered by the call option. Therefore, the option should be exercised only at expiry date T, as for its European counterpart. In what follows, we will consider American put options only.

© The Author(s), under exclusive license to Springer Nature Singapore Pte Ltd. 2020   35
S. Wang, *The Fitted Finite Volume and Power Penalty Methods*
for Option Pricing, SpringerBriefs in Mathematical Methods,
https://doi.org/10.1007/978-981-15-9558-5_2

2.1.1 The Differential LCP

Consider an American put option (put) with strike price K, maturity T and payoff $V^*(S) := \max(K - S, 0)$, under the market assumptions in Sect. 1.2. Let us consider from the option holder's point of view the hedging problem: for every American put on a stock you buy, from how many shares of the stock you can neutralize the risk. Equivalently, let us consider a riskless portfolio Π consisting of one put option and δ shares of the underlying stock. At t, the value of the portfolio is $\Pi = \delta S + V(S, t)$, where S denotes the price of the stock and V the value of the option. Using the same argument in Sect. 1.2.1, we have that S and V should satisfy (1.1.4) and

$$d\Pi = \left[\mu S \delta + \left(\frac{\partial V}{\partial t} + \frac{1}{2}\sigma^2 S^2 \frac{\partial^2 V}{\partial S^2} + \mu S \frac{\partial V}{\partial S} \right) \right] dt + \sigma S \left(\delta + \frac{\partial V}{\partial S} \right) dW.$$

Note that $\frac{\partial V}{\partial S} \leq 0$ for a put. Since we expect Π is riskless, the coefficient of dW should vanish, implying $\delta = -\frac{\partial V}{\partial S}$. Thus, the above equality becomes

$$d\Pi = \left(\frac{\partial V}{\partial t} + \frac{1}{2}\sigma^2 S^2 \frac{\partial^2 V}{\partial S^2} \right) dt. \tag{2.1.1}$$

We expect that, if the option is exercised at the optimal time, this portfolio has the riskless return rate r, that is $d\Pi = r\Pi dt = r(-S\frac{\partial V}{\partial S} + V)dt$. However, if the option is exercised not optimally, Π should have a return rate less than r. Thus, from (2.1.1) and the above analysis we have $\left(\frac{\partial V}{\partial t} + \frac{1}{2}\sigma^2 S^2 \frac{\partial^2 V}{\partial S^2} \right) \leq r \left(-S\frac{\partial V}{\partial S} + V \right)$, or

$$\frac{\partial V}{\partial t} + \frac{1}{2}\sigma^2 S^2 \frac{\partial^2 V}{\partial S^2} + rS\frac{\partial V}{\partial S} - rV \leq 0. \tag{2.1.2}$$

This inequality becomes an equation if the option is exercised at the optimal time.

Furthermore, at any t, if the spot price $S < K$, the holder of the American put can buy a share of the stock from the share market and exercise the put so that the holder makes a profit of $K - S$. On the other hand, $V(S, t) \geq 0$. Thus, combining these two cases, we have $V(S, t) \geq V^*(S) := \max(0, K - S)$, where $V^*(S)$ is called the intrinsic value of the American put option.

From the above analysis, we see that there is an optimal exercise curve $S_{\mathrm{opt}}(t)$ such that when $S(t) > S_{\mathrm{opt}}(t)$ the option should be held and (2.1.2) becomes an equation, as Π has the return rate r. When $S(t) < S_{\mathrm{opt}}(t)$, it should be exercised, as otherwise the return rate of Π is smaller than r and the option value falls below $V^*(S)$. Combining these with (2.1.2), we see $V(S, t)$ satisfies

$$-\left(\frac{\partial V}{\partial t} + \frac{1}{2}\sigma^2 S^2 \frac{\partial^2 V}{\partial S^2} + rS\frac{\partial V}{\partial S} - rV\right) \geq 0, \quad , V - V^* \geq 0, \qquad (2.1.3)$$

$$-\left(\frac{\partial V}{\partial t} + \frac{1}{2}\sigma^2 S^2 \frac{\partial^2 V}{\partial S^2} + rS\frac{\partial V}{\partial S} - rV\right)(V - V^*) = 0 \qquad (2.1.4)$$

for $(S, t) \in (0, \infty) \times [0, T)$ with $V(S, T) = V^*(S)$. This is an LCP.

2.1.2 The Variational Inequality

The LCP (2.1.3)–(2.1.4) is defined on an infinite domain. In computation, we usually truncate this infinite domain into $I \times [0, T)$, where $I = (0, S_{\max})$ for an $S_{\max} \gg K$. Let us consider the following general LCP governing a American put option valuation:

$$LV(S, t) \geq 0, \quad V(S, t) \geq V^*(S), \quad LV(S, t) \cdot (V(S, t) - V^*(S)) = 0 \quad (2.1.5)$$

for $(S, t) \in I \times [0, T)$ with the payoff condition $V(S, T) = V^*(S)$, where L is the differential operator defined in (1.2.3). The above LCP is equivalent to the Hamilton-Jacobi-Bellman equation $\min\{LV, V - V^*\} = 0$ for $(S, t) \in I \times [0, T)$.

We now define boundary conditions for $V(S, t)$. When $S_{\max}$ is sufficiently large, we have $V(S_{\max}, t) = 0$. When $S \to 0^+$, it is obvious that the option should be exercised and the payoff is $V(0, t) = V^*(0) = K$. Clearly, (2.1.5) and these boundary and payoff conditions determine the value of an American put with variable (but not stochastic) volatility, interest rate and dividend, which contain (2.1.3)–(2.1.4) as a special case.

Let $V_0(S) = \left(1 - \frac{S}{S_{\max}}\right) K$, a special case of (1.2.11). Clearly, $V_0(S)$ satisfies the same boundary conditions as $V(S, t)$ does. Introduce a new variable

$$u(S, t) = -e^{\beta t}(V(S, t) - V_0(S)), \qquad (2.1.6)$$

where $\beta > 0$ is a constant to be determined later. Using this transformation and the operator $\mathscr{L}$ defined in (1.2.12), we rewrite (2.1.5) as the following LCP:

$$\mathscr{L}u \leq f, \quad u - u^* \leq 0, \quad (\mathscr{L}u - f)(u - u^*) = 0, \quad (S, t) \in I \times [0, T), \quad (2.1.7)$$

where $f(t) = e^{\beta t} L V_0(S)$ and $u^*(x) = e^{\beta t}(V_0(S) - V^*(S))$. It is easily seen

$$u^*(S, t) = e^{\beta t}(V_0(S) - V^*(S)) = \begin{cases} e^{\beta t}\left(1 - \frac{K}{S_{\max}}\right) S, \ 0 \leq S \leq K, \\ e^{\beta t}\left(1 - \frac{S}{S_{\max}}\right) K, \ K < S \leq S_{\max}. \end{cases} \quad (2.1.8)$$

From (2.1.6) we see that the boundary and payoff conditions for (2.1.7) are

$$u(0, t) = 0 = u(S_{\max}, t) \, t \in [0, T) \text{ and } u(S, T) = u^*(S, T).$$

Let $\mathcal{K} = \{v \in H^1_{0,w}(I) : v \leq u^*\}$, where $H^1_{0,w}(I)$ is the Sobolev space defined in Sect. 1.2.3. It is easy to verify that $\mathcal{K}$ is a convex and closed subset of $H^1_{0,w}(I)$. Using $\mathcal{K}$, we define the following problem.

Problem 2.1.1 *Find $u(t) \in \mathcal{K}$ such that, for all $v \in \mathcal{K}$,*

$$\left(-\frac{\partial u(t)}{\partial t}, v - u(t) \right) + A(u(t), v - u(t); t) \geq (f(t), v - u(t)) \tag{2.1.9}$$

a.e. in $[0, T)$, where $A(u, v; t) = \left(aS^2 \frac{\partial u}{\partial S} + bSu, \frac{\partial v}{\partial S} \right) + (cu, v)$ as in (1.2.18).

For this variational inequality problem, we have the following theorem.

Theorem 2.1.1 *Problem 2.1.1 is the variational form corresponding to the linear complementarity problem (2.1.7).*

Proof Note that $w - u^*(t) \leq 0$ a.e. on $I \times [0, T)$ for any $w \in \mathcal{K}$. Multiplying both sides of $\mathcal{L}u \leq f$ in (2.1.7) by $w - u^*$ for an arbitrary $w \in \mathcal{K}$ and integrating the second term by parts as in the proof of Theorem 1.2.1, we obtain

$$\left(-\frac{\partial u}{\partial t}, w - u^* \right) + A(u, w - u^*; t) \geq (f, w - u^*), \, t \in [0, T) \text{ a.e.} \tag{2.1.10}$$

Since $\mathcal{K}$ is a convex subset of $H^1_{0,w}(I)$, we may write w as $w = \theta v + (1 - \theta)u(t)$, where $v \in \mathcal{K}$ and $\theta \in [0, 1]$. Therefore, (2.1.10) becomes

$$\left(-\frac{\partial u(t)}{\partial t}, u(t) - u^*(t) \right) + \left(-\frac{\partial u(t)}{\partial t}, \theta(v - u(t)) \right) + A(u(t), u(t) - u^*(t); t)$$
$$+ A(u(t), \theta(v - u(t)); t) \geq (f(t), u(t) - u^*(t)) + (f(t), \theta(v - u(t))). \tag{2.1.11}$$

Integrating $(\mathcal{L}u - f)(u - u^*) = 0$ in (2.1.7) by parts, we have

$$\left(-\frac{\partial u(t)}{\partial t}, u(t) - u^*(t) \right) + A(u(t), u(t) - u^*(t); t) = (f(t), u(t) - u^*(t)).$$

The difference between (2.1.11) and the above equality is

$$\left(-\frac{\partial u(t)}{\partial t}, \theta(v - u(t)) \right) + A(u(t), \theta(v - u(t)); t) \geq (f(t), \theta(v - u(t))).$$

Since $\theta \geq 0$, eliminating it from the above inequality we have (2.1.9).   □

For Problem 2.1.1 we have the following theorem.

Theorem 2.1.2 *Let Assumption (1.2.1) be fulfilled. Then, Problem 2.1.1 has a unique solution in the convex set $\mathcal{K}$ defined above.*

Proof When Assumption (1.2.1) is satisfied by σ, r, D and β in (2.1.6), it has been shown in the proof of Theorem 1.2.1 that, for any $v, w \in H^1_{0,w}(I)$, there exist positive constants C and M, independent of v and w, such that

$$A(v, v; t) \geq C\|v\|^2_{1,w}, \quad A(v, w; t) \leq M\|v\|_{1,w}\|w\|_{1,w}, \tag{2.1.12}$$

i.e., $A(\cdot, \cdot; t)$ is coercive and Litschitz continuous with respect to any of its two variables. Therefore, the unique solbability of Problem 2.1.1 is just a consequence of (2.1.12) and [8, Theorem 1.33], in which the unique solvability for a general variational inequality problem is established. $\qquad\square$

2.2 The Penalty Method and Its Convergence Analysis

Numerical solution of variational inequalities of the form (2.1.9) (or (2.1.3)–(2.1.4)) has been discussed extensively in the open literature (e.g., [6, 8]). Numerical methods for pricing American options have been proposed by numerous authors (e.g., [1, 4, 7, 10–12]). Since (2.1.3)–(2.1.4) is closely related to a constrained optimization problem, optimization techniques are expected to be used for it. In recent years, linear penalty methods have been used successfully for LCPs including American option pricing problems (e.g., [3, 5, 6]). A power penalty method for (2.1.3)–(2.1.4) has been proposed and analyzed in [13].

2.2.1 The Power Penalty Equation and Its Solvability

Let us consider the following simple constrained optimization or obstacle problem: find $u \in H^1_0(I)$ such that $u = \arg\inf_{v \in H^1_0(I), v \leq 0} \int_I \left(\frac{1}{2}|\nabla v|^2 - fv\right) dS$. Its solution u satisfies [9]: $(\nabla u, \nabla(v - u)) \geq f(v - u), \forall v \in \{w \in H^1_0(I) : w \leq 0\}$.

The above constrained optimization problem can be approximated by

$$\inf_{v \in H^1_0(I)} \int_I \left(\frac{1}{2}|\nabla v|^2 - fv + \frac{\lambda}{1 + 1/\kappa}[v]^{1+1/\kappa}_+\right) dS,$$

where $[z]_+ = \max\{z, 0\}$ for any function z, and $\kappa > 0$ and $\lambda > 1$ are positive constants. From Calculus of Variation we see that the optimality condition (Euler–Lagrange equation) for this optimization problem is $-\nabla^2 v + \lambda[v]^{1/\kappa}_+ = f$.

Motivated by the above example, we propose to approximate (2.1.9)/(2.1.7): by the following nonlinear equation system:

$$\mathscr{L}u_\lambda(S, t) + \lambda[u_\lambda(S, t) - u^*(S, t)]^{1/\kappa}_+ = f(S, t), \quad (S, t) \in I \times [0, T), \tag{2.2.1}$$

$$u_\lambda(0, t) = 0 = u_\lambda(S_{\max}, t) \quad \text{and} \quad u_\lambda(S, T) = u^*(S, T), \tag{2.2.2}$$

where $\lambda > 1$ and $\kappa > 0$ are parameters. In (2.2.1), $\lambda[u_\lambda(S, t) - u^*(S, t)]_+^{1/\kappa}$ is the penalty term which penalizes the positive part of $u_\lambda - u^*$. Using the argument for deducing (1.2.17), we have the variational problem corresponding to (2.2.1)–(2.2.2) as follows.

Problem 2.2.1 *Find $u_\lambda(t) \in H_{0,w}^1(I)$ such that, for all $v \in H_{0,w}^1(I)$,*

$$\left(-\frac{\partial u_\lambda}{\partial t}, v\right) + A(u_\lambda, v; t) + \lambda\left([u_\lambda - u^*]_+^{1/\kappa}, v\right) = (f, v), \quad \text{a.e. in } (0, T).$$

$$(2.2.3)$$

For any Hilbert space $H(I)$, we let $L^p(0, T; H(I))$ denote the space defined by

$$L^p(0, T; H(I)) = \{v(\cdot, t) : v(\cdot, t) \in H(I) \text{ a.e. in } (0, T); \|v(\cdot, t)\|_H \in L^p((0, T))\},$$

where $1 \le p \le \infty$ and $\| \cdot \|_H$ denotes the natural norm on $H(I)$. The norm on this space is denoted by $\| \cdot \|_{L^p(0,T;H)}$, i.e.,

$$\|v\|_{L^p(0,T;H(I))} = \left(\int_0^T \|v(\cdot, t)\|_H^p dt\right)^{1/p}. \qquad (2.2.4)$$

Clearly, $L^p(0, T; L^p(I)) = L^p(I \times (0, T)) = L^p(\Omega)$.

Using this space, we have the following theorem.

Theorem 2.2.1 *Problem 2.2.1 has a unique solution in $H_{0,w}^1(I)$.*

Proof First, from (1.2.11), we see $f(S, t) = e^{\beta t} L V_0$ is sufficiently smooth in (S, t). We now prove this theorem by showing that the variational form of the nonlinear operator on the LHS of (2.2.1) is strongly monotone and continuous. For any $v_1(t), v_2(t) \in H_{0,w}^1(I)$ with $v_1(T) = v_2(T) = u^*(S, T)$, let $e = v_1 - v_2$. Then, using integration by parts, we have

$$(\mathcal{L}e, e) + \lambda\left([v_1 - u^*]_+^{1/\kappa} - [v_2 - u^*]_+^{1/\kappa}, e\right)$$

$$= \left(-\frac{\partial e}{\partial t}, e\right) + A(e, e; t) + \lambda([v_1 - u^*]_+^{1/\kappa} - [v_2 - u^*]_+^{1/\kappa}, e). \qquad (2.2.5)$$

By definition, $[v]_+^{1/\kappa} = (\max\{v, 0\})^{1/\kappa}$. Clearly, $[v]_+^{1/\kappa}$ is monotone in v, and so

$$\lambda\left([v_1 - u^*]_+^{1/\kappa} - [v_2 - u^*]_+^{1/\kappa}, e\right) \ge 0.$$

Integrating both sides of (2.2.5) from 0 to T and using the above inequality and (2.1.12), we have

$$\int_0^T \left[(\mathscr{L}e(\tau), e(\tau)) + \lambda \left([v_1(\tau) - u^*(\tau)]_+^{1/\kappa} - [v_2(\tau) - u^*(\tau)]_+^{1/\kappa}, e(\tau) \right) \right] d\tau$$

$$= \int_0^T \left(-\frac{\partial e(\tau)}{\partial \tau}, e(\tau) \right) d\tau + \int_0^T \left[A(e(\tau), e(\tau)) + \lambda([v_1(\tau) - u^*(\tau)]_+^{1/\kappa} \right.$$

$$\left. - [v_2(\tau) - u^*(\tau)]_+^{1/\kappa}, e(\tau)) \right] d\tau \geq \int_0^T \left(-\frac{\partial(\tau)}{\partial t}, e(\tau) \right) d\tau + C \int_0^T \|e(\tau)\|_{1,w}^2 d\tau.$$

$$(2.2.6)$$

For any $t \in (0, T)$, using integrating by parts we have

$$\int_t^T \left(-\frac{\partial e(\tau)}{\partial \tau}, e(\tau) \right) d\tau = (e(t), e(t)) - \int_t^T \left(-\frac{\partial e(\tau)}{\partial \tau}, e(\tau) \right) d\tau,$$

because $e(T) = 0$. From this, it follows that

$$\int_t^T \left(-\frac{\partial e(\tau)}{\partial \tau}, e(\tau) \right) d\tau = \frac{1}{2}(e(t), e(t)) \geq 0. \qquad (2.2.7)$$

Therefore, from (2.2.6), (2.2.7) and (2.2.4), we get

$$\int_0^T \left[(\mathscr{L}e, e) + \lambda \left([v_1 - u^*]_+^{1/\kappa} - [v_2 - u^*]_+^{1/\kappa}, e \right) \right] d\tau \geq C \|e\|_{L^2(0,T;H_{0,w}^1(I))}^2$$

with $e = v_1 - v_2$. This implies that the operator on the right-hand side of (2.2.1) is strongly monotone.

Moreover, for any $v, w \in L^2(0, T; H_{0,w}^1(I))$, it is easy to show using the 2nd inequality in (2.1.12) and a standard argument that

$$\int_0^T [A(v, w; t)] \, dt \leq C \|v\|_{L^2(0,T;H^1 0,w(I))} \|w\|_{L^2(0,T;H^1 0,w(I))}.$$

Also, it is easily seen that $([v - u^*]_+^{1/\kappa}, w)$ is also continuous in both v and w. Therefore, using the result in [8, p. 37], we have that Problem 2.2.1 is uniquely solvable. $\qquad\Box$

2.2.2 Convergence Analysis

Since Problems 2.1.1 and 2.2.1 are not equivalent, it is necessary to prove that the solution to Problem 2.2.1 converges to that of Problem 2.1.1 as $\lambda \to \infty$ in a proper norm. We shall prove this in two different stages. In Stage 1, we establish an error bound for $[u_\lambda - u^*]_+$. In Stage 2, we use this bound to derive a bound for $u_\lambda - u$. We begin this analysis with the following lemma.

Lemma 2.2.1 *Let u_λ be the solution to Problem 2.2.1. If $u_\lambda \in L^p(\Omega)$, then there exists a positive constant C, independent of u_λ and λ, such that*

$$\|[u_\lambda - u^*]_+\|_{L^p(\Omega)} \le \frac{C}{\lambda^\kappa}, \qquad (2.2.8)$$

$$\|[u_\lambda - u^*]_+\|_{L^\infty(0,T;L^2(I))} + \|[u_\lambda - u^*]_+\|_{L^2(0,T;H^1_{0,w}(I))} \le \frac{C}{\lambda^{\kappa/2}}, \qquad (2.2.9)$$

where $p = 1 + 1/\kappa$.

Proof Let C be a generic positive constant, independent of u_λ and λ. In what follows, we use this single constant C in our bound estimations. To simplify the notation, we introduce $\phi(\cdot, t) = [u_\lambda(\cdot, t) - u^*(\cdot)]_+ \in H^1_{0,w}(I)$ a.e. in $(0, T)$.

Since v in (2.2.3) is arbitrary, replacing it with ϕ gives

$$\left(-\frac{\partial u_\lambda}{\partial t}, \phi\right) + A(u_\lambda, \phi; t) + \lambda(\phi^{1/\kappa}, \phi) = (f, \phi) \quad \text{a.e. in } (0, T).$$

Subtracting $-(\frac{\partial u^*}{\partial t}, \phi) + A(u^*, \phi; t)$ from both sides of the above equation gives

$$\left(-\frac{\partial(u_\lambda - u^*)}{\partial t}, \phi\right) + A(u_\lambda - u^*, \phi; t) + \lambda(\phi^{1/\kappa}, \phi) = (f, \phi) + \left(\frac{\partial u^*}{\partial t}, \phi\right) - A(u^*, \phi; t).$$

Integrating this from t to T and using (2.1.8) and Hölder inequality yield

$$\int_t^T \left(-\frac{\partial(u_\lambda - u^*)}{\partial \tau}, \phi\right) d\tau + \int_t^T A(u_\lambda - u^*, \phi; \tau) d\tau + \lambda \int_t^T (\phi^{1/\kappa}, \phi)$$

$$= \int_t^T (f, \phi) d\tau + \int_t^T \left(\frac{\partial u^*}{\partial t}, \phi\right) d\tau - \int_t^T A(u^*, \phi; \tau) d\tau$$

$$\le \left(\int_t^T \|f(\tau)\|^q_{L^q(I)} d\tau\right)^{1/q} \left(\int_t^T \|\phi(\tau)\|^p_{L^p(I)} d\tau\right)^{1/p}$$

$$+ \beta \int_t^T e^{\beta\tau} (V_0 - V^*, \phi(\tau)) d\tau - \int_t^T A(u^*(\tau), \phi(\tau); \tau) d\tau,$$

where $q = 1 + \kappa$ so that $1/p + 1/q = 1$.

From the definition of ϕ, we see $\frac{\partial(u_\lambda - u^*)}{\partial t} \cdot \phi = \frac{\partial \phi}{\partial t} \cdot \phi$. Thus, from the above estimate, (2.2.7) and the 1st inequality in (2.1.12) we get

$$\frac{1}{2}(\phi, \phi) + C \int_t^T \|\phi\|^2_{H^1_{0,w}(I)} d\tau + \lambda \int_t^T \|\phi\|^p_{L^p(I)} d\tau \le C \left(\int_t^T \|\phi\|^p_{L^p(I)} d\tau\right)^{1/p}$$

$$+ \beta \int_t^T e^{\beta\tau} (V_0 - V^*, \phi) d\tau - \int_t^T A(u^*, \phi; \tau) \tau. \qquad (2.2.10)$$

Let us consider the last two integrals in (2.2.10). From (2.1.8), we see that $|V_0(S) - V^*(S)| \leq (1 - K/S_{\max})K$ for $S \in [0, S_{\max}]$. Using this bound and Hölder inequality we have the following estimate:

$$\int_t^T e^{\beta \tau} \left(V_0 - V^*, \phi\right) d\tau \leq C \int_t^T \int_0^{S_{\max}} \phi \, dS \, d\tau \leq C \left(\int_t^T \|\phi\|_{L^p(I)}^p d\tau\right)^{1/p}.$$
(2.2.11)

From (1.2.18), we see that the integrand of the last term in (2.2.10) is

$$- A(u^*, \phi; \tau) = -\left(aS^2 \frac{\partial u^*}{\partial S} + bSu^*, \frac{\partial \phi}{\partial S}\right) + (cu^*, \phi).$$
(2.2.12)

Using (2.1.8), we have that

$$\frac{\partial u^*}{\partial S} = \begin{cases} e^{\beta t}(1 - K/S_{\max}), & S \in (0, K) \\ -e^{\beta t} K/S_{\max}, & S \in (K, S_{\max}). \end{cases}$$
(2.2.13)

Therefore, integrating by parts and using $u^*(0, t) = 0 = u^*(S_{\max}, t)$, we obtain

$$-\int_0^{S_{\max}} a(\tau) S^2 \frac{\partial u^*(S, \tau)}{\partial S} \cdot \frac{\partial \phi(S, \tau)}{\partial S} dS$$

$$= -e^{\beta \tau}\left(1 - \frac{K}{S}\right) \int_0^K aS^2 \frac{\partial \phi}{\partial S} dS + e^{\beta \tau} \frac{K}{S_{\max}} \int_K^{S_{\max}} aS^2 \frac{\partial \phi}{\partial S} dS$$

$$= -e^{\beta \tau} aK^2 \phi(K, \tau) + e^{\beta \tau}\left(1 - \frac{K}{S}\right) \int_0^K 2aS\phi \, dS$$

$$+ e^{\beta \tau} \frac{K}{S_{\max}} \int_K^{S_{\max}} 2aS\phi \, dS \leq C \int_0^{S_{\max}} \phi(S, \tau) dS$$

for $\tau \in (0, T)$, because $a(\tau) = \frac{1}{2}\sigma^2(\tau) > 0$ and $\phi \geq 0$. Similarly, since $u^* \in H_0^1(I)$ and $b, c \in L^\infty(I)$, using integration by parts and (2.2.13) we have

$$-\left(bSu^*, \frac{\partial \phi}{\partial S}\right) + (cu^*, \phi) = \left(b(u^* + S\frac{\partial u^*}{\partial S}), \phi\right) + (cu^*, \phi) \leq C \int_0^{S_{\max}} \phi \, dS.$$

Integrating from t to T and using the above two estimates, we have from (2.2.12)

$$-\int_t^T A(u^*, \phi; \tau) d\tau \leq C \int_t^T \int_0^{S_{\max}} \phi \, dS \, d\tau \leq C \left(\int_t^T \|\phi\|_{L^p(I)}^p d\tau\right)^{1/p}.$$

Combining the above inequality and (2.2.11) with (2.2.10), we obtain

$$\frac{1}{2}(\phi, \phi) + \int_t^T \|\phi\|_{H_{0,w}^1(I)}^2 d\tau + \lambda \int_t^T \|\phi\|_{L^p(I)}^p d\tau \le C \left(\int_t^T \|\phi\|_{L^p(I)}^p d\tau \right)^{1/p}$$

(2.2.14)

for all $t \in (0, T)$, which implies

$$\lambda \int_t^T \|\phi\|_{L^p(I)}^p d\tau \le C \left(\int_t^T \|\phi\|_{L^p(I)}^p d\tau \right)^{1/p}, \text{ or } \left(\int_t^T \|\phi\|_{L^p(I)}^p d\tau \right)^{1-1/p} \le \frac{C}{\lambda}.$$

Taking the $(p-1)$th root on both sides of the above estimate gives

$$\left(\int_t^T \|\phi(\tau)\|_{L^p(I)}^p d\tau \right)^{1/p} \le \frac{C}{\lambda^{1/(p-1)}} = \frac{C}{\lambda^\kappa}, \qquad (2.2.15)$$

since $p = 1 + 1/\kappa$. Thus, we have proved (2.2.8).

From (2.2.14) and (2.2.15), we have

$$\frac{1}{2}(\phi(t), \phi(t)) + \int_t^T \|\phi(\tau)\|_A^2 d\tau \le C \left(\int_t^T \|\phi(\tau)\|_{L^p(I)}^p d\tau \right)^{1/p} \le \frac{C}{\lambda^\kappa},$$

from which it follows that $(\phi(t), \phi(t))^{1/2} + \left(\int_t^T \|\phi(\tau)\|_A^2 d\tau \right)^{1/2} \le \frac{C}{\lambda^{\kappa/2}}$ for all $t \in (0, T)$. Thus, we have proved (2.2.9). □

Using the results obtained in Lemma 2.2.1 we are able to establish the rate of convergence for u_λ in the following theorem.

Theorem 2.2.2 *Let $\frac{\partial u}{\partial t} \in L^{\kappa+1}(\Omega)$ and the assumptions in Lemma 2.2.1 be fulfilled. Then, the solutions u and u_λ to Problems 2.1.1 and 2.2.1 respectively. satisfy*

$$\|u - u_\lambda\|_{L^\infty(0,T;L^2(I))} + \|u - u_\lambda\|_{L^2(0,T;H_{0,w}^1(I))} \le \frac{C}{\lambda^{\kappa/2}}, \qquad (2.2.16)$$

where C is a positive constant, independent of u, u_λ and λ.

Proof We continue to use $\phi(\cdot, t)$ introduced in the proof of Lemma 2.2.1.

The term $u - u_\lambda$ can be decomposed as

$$u - u_\lambda = u - u^* - (u_\lambda - u^*) = u - u^* + [u_\lambda - u^*]_- - [u_\lambda - u^*]_+ =: r_\lambda - \phi,$$

(2.2.17)

where $[u_\lambda - u^*]_- = -\min\{u_\lambda - u^*, 0\}$ and $r_\lambda = u - u^* + [u_\lambda - u^*]_-$. From the definition of ϕ, we see $(\phi^\alpha, [u_\lambda - u^*]_-) = [u - u^*]_+^\alpha [u_\lambda - u^*]_- = 0$ for any $\alpha > 0$.

To establish an upper bound for $u - u_\lambda$, we need only to find an upper bound for r_λ, as that for ϕ is given in Lemma 2.2.1. Setting $v = u - r_\lambda$ in (2.1.9) and $v = r_\lambda$ in (2.2.3) respectively, we have

$$\left(-\frac{\partial u}{\partial t}, -r_\lambda\right) + A(u, -r_\lambda; t) \geq (f, -r_\lambda),$$

$$\left(-\frac{\partial u_\lambda}{\partial t}, r_\lambda\right) + A(u_\lambda, r_\lambda; t) + \lambda(\phi^{1/\kappa}, r_\lambda) = (f, r_\lambda).$$

Adding up the above inequality and equality gives

$$\left(-\frac{\partial(u_\lambda - u)}{\partial t}, r_\lambda\right) + A(u_\lambda - u, r_\lambda; t) + \lambda(\phi^{1/\kappa}, r_\lambda) \geq 0. \tag{2.2.18}$$

Using the definition of r_λ, we have

$$(\phi^{1/\kappa}, r_\lambda) = (\phi^{1/\kappa}, u - u^* + [u_\lambda - u^*]_-) = (\phi^{1/\kappa}, u - u^*) \leq 0, \tag{2.2.19}$$

since $\phi \geq 0$, $u - u^* \leq 0$ and $\phi[u_\lambda - u^*]_- = 0$. From (2.2.18) and (2.2.19), we get $\left(-\frac{\partial(u-u_\lambda)}{\partial t}, r_\lambda\right) + A(u - u_\lambda, r_\lambda; t) \leq 0$, and using this inequality and the decomposition of $u - u_\lambda$ in (2.2.17), we have

$$\left(-\frac{\partial r_\lambda}{\partial t}, r_\lambda\right) + A(r_\lambda, r_\lambda; t) \leq \left(-\frac{\partial \phi}{\partial t}, r_\lambda\right) + A(\phi, r_\lambda; t).$$

Integrating both sides of the above estimate from t to T and using (2.2.7) and Cauchy–Schwarz inequality, we obtain

$$\frac{1}{2}(r_\lambda(t), r_\lambda(t)) + \int_t^T A(r_\lambda(\tau), r_\lambda(\tau); \tau)d\tau$$

$$\leq \int_t^T \left(-\frac{\partial \phi(\tau)}{\partial \tau}, r_\lambda(\tau)\right) d\tau + \int_t^T A(\phi(\tau), r_\lambda(\tau); \tau)d\tau$$

$$\leq (\phi(t), r_\lambda(t)) + \int_t^T \left(\phi(\tau), \frac{\partial r_\lambda(\tau)}{\partial \tau}\right) d\tau + \int_t^T A(\phi(\tau), r_\lambda(\tau); \tau)d\tau$$

$$\leq \|\phi\|_{L^\infty(0,T;L^2(I))} \|r_\lambda\|_{L^\infty(0,T;L^2(I))} + C\|\phi\|_{L^2(0,T;H^1_{0,w}(I))} \|r_\lambda\|_{L^2(0,T;H^1_{0,w}(I))}$$

$$+ \int_t^T \left(\phi(\tau), \frac{\partial r_\lambda(\tau)}{\partial t}\right) d\tau, \quad t \in (0, T). \tag{2.2.20}$$

From (2.1.8) and the definition of r_λ, we have

$$\phi(\tau)\frac{\partial r_\lambda(\tau)}{\partial \tau} = \phi(\tau)\left(\frac{\partial u(\tau)}{\partial \tau} - \frac{\partial u^*(\tau)}{\partial \tau}\right) = \phi(\tau)\frac{\partial u(\tau)}{\partial \tau} - \phi(\tau)\beta e^{\beta\tau}(V_0 - V^*).$$

Thus, using (2.2.8) we obtain

$$\int_t^T \left(\phi(\tau), \frac{\partial r_\lambda(\tau)}{\partial \tau}\right) d\tau = \int_t^T \left(\phi(\tau), \frac{\partial u(\tau)}{\partial \tau}\right) d\tau - \beta \int_t^T e^{\beta \tau} \left(\phi(\tau), V_0 - V^*\right) d\tau$$

$$\le C\|\phi\|_{L^p(\Omega)} \left(\left\|\frac{\partial u}{\partial t}\right\|_{L^q(\Omega)} + \|V_0 - V^*\|_{L^q(\Omega)}\right) \le \frac{C}{\lambda^\kappa},$$

since $\frac{\partial u}{\partial t} \in L^{\kappa+1}(\Omega)$, where $p = 1 + 1/\kappa$ and $q = \kappa + 1$. Substituting the above upper bound into (2.2.20) and using (2.2.9), we obtain

$$\frac{1}{2}(r_\lambda(t), r_\lambda(t)) + \int_t^T A(r_\lambda(\tau), r_\lambda(\tau); \tau) d\tau \cdot$$

$$\le \|\phi\|_{L^\infty(0,T;L^2(I))} \|r_\lambda\|_{L^\infty(0,T;L^2(I))} + C\|\phi\|_{L^2(0,T;H^1_{0,w}(I))} \|r_\lambda\|_{L^2(0,T;H^1_{0,w}(I))} + \frac{C}{\lambda^\kappa}$$

$$\le C\left(\|\phi\|_{L^\infty(0,T;L^2(I))} + \|\phi\|_{L^2(0,T;H^1_{0,w}(I))}\right)\left(\|r_\lambda\|_{L^\infty(0,T;L^2(I))} + \|r_\lambda\|_{L^2(0,T;H^1_{0,w}(I))}\right)$$

$$+ C\lambda^{-\kappa} \le C\left[\lambda^{-\kappa/2}\left(\|r_\lambda\|_{L^\infty(0,T;L^2(I))} + \|r_\lambda\|_{L^2(0,T;H^1_{0,w}(I))}\right) + \lambda^{-\kappa}\right].$$

On the other hand, from (2.1.12), (2.2.7), (2.2.4) and the above estimate, we have

$$\left(\|r_\lambda\|_{L^\infty(0,T;L^2(I))} + \|r_\lambda\|_{L^2(0,T;H^1_{0,w}(I))}\right)^2 \le \frac{C}{2}\|r_\lambda\|^2_{L^\infty(0,T;L^2(I))} + C\|r_\lambda\|^2_{L^2(0,T;H^1_{0,w}(I))}$$

$$\le \frac{C}{2}(r_\lambda(t), r_\lambda(t)) + C\int_t^T A(r_\lambda(\tau), r_\lambda(\tau); \tau) d\tau$$

$$\le C\left[\lambda^{-\kappa/2}\left(\|r_\lambda\|_{L^\infty(0,T;L^2(I))} + \|r_\lambda\|_{L^2(0,T;H^1_{0,w}(I))}\right) + \lambda^{-\kappa}\right]. \qquad (2.2.21)$$

This is of the form $y^2 \le C\rho^{1/2}y + C\rho$ which can be rewritten as $(y - \frac{1}{2}C\rho^{1/2})^2 \le \left(C + \frac{C^2}{4}\right)\rho$. From this inequality, we have $y \le C\rho^{1/2}$. (Recall that $C > 0$ is a generic constant.) Thus, applying this analysis to (2.2.21) yields

$$\|r_\lambda\|_{L^\infty(0,T;L^2(I))} + \|r_\lambda\|_{L^2(0,T;H^1_0(I))} \le \frac{C}{\lambda^{\kappa/2}}.$$

Using the triangle inequality, the above estimate and (2.2.9), we have from (2.2.17),

$$\|u - u_\lambda\|_{L^\infty(0,T;L^2(I))} + \|u - u_\lambda\|_{L^2(0,T;H^1_{0,w}(I))} \le \|r_\lambda\|_{L^\infty(0,T;L^2(I))} + \|r_\lambda\|_{L^2(0,T;H^1_{0,w}(I))}$$

$$+ \left(\|\phi\|_{L^\infty(0,T;L^2(I))} + \|\phi\|_{L^2(0,T;H^1_{0,w}(I))}\right) \le \frac{C}{\lambda^{\kappa/2}}.$$

This is (2.2.16), and thus the theorem is proved.   □

Remark 2.2.1 The assumption $\frac{\partial u}{\partial t} \in L^{\kappa+1}$ in Theorem 2.2.2 is not restrictive. Any function which has a bounded derivative w.r.t. t satisfies this condition. A function v with a singular t-derivative, $\partial v/\partial t$, is also in $L^{\kappa+1}(\Omega)$ if $\int_I |\frac{\partial v}{\partial t}|^{\kappa+1} dS < \infty$.

2.3 Numerical Solution of the Penalty Equation

We now consider the numerical solution of (2.2.1) which is a nonlinear parabolic PDE. We discretize it using the FVM in Sect. 1.3 and use a Newton's method for the resulting nonlinear system.

2.3.1 Discretization

To simplify notation, we suppress the subscript λ of u_λ, but bear in mind u is a solution to Problem 2.2.1. In the following discussion, we use the notation defined in Sect. 1.3.

Let the interval $I = (0, S_{\max})$ be divided into N sub-intervals $I_i := (S_i, S_{i+1}), i = 0, 1, \ldots, N - 1$ with $0 = S_0 < S_1 < \cdots < S_N = S_{\max}$. For each $i = 0, 1, \ldots, N - 1$, we put $h_i = S_{i+1} - S_i$ and $h = \max_{0 \le i \le N-1} h_i$. We also let $S_{i-1/2} = (S_{i-1} + S_i)/2$ and $S_{i+1/2} = (S_i + S_{i+1})/2$ for each $i = 1, 2, \ldots, N - 1$. These mid-points form a second partition of $(0, S_{\max})$ if we define $S_{-1/2} = S_0$ and $S_{N+1/2} = S_N$.

For any $i = 1, 2, \ldots, N - 1$, integrating (2.2.1) over $(S_{i-1/2}, S_{i+1/2})$ we have

$$-\int_{S_{i-\frac{1}{2}}}^{S_{i+\frac{1}{2}}} \frac{\partial u}{\partial t} dx - [S\rho(u)]_{S_{i-\frac{1}{2}}}^{S_{i+\frac{1}{2}}} + \int_{S_{i-\frac{1}{2}}}^{S_{i+\frac{1}{2}}} cu\,dS + \lambda \int_{S_{i-\frac{1}{2}}}^{S_{i+\frac{1}{2}}} [u - u^*]_+^{1/\kappa} dS = \int_{S_{i-\frac{1}{2}}}^{S_{i+\frac{1}{2}}} f\,dS,$$

where $\rho(u)$ is the flux defined in (1.3.2). Applying a 1-point quadrature rule to the first, third, fourth and last terms, we obtain from the above

$$-\frac{\partial u_i}{\partial t} l_i - \left[S_{i+\frac{1}{2}} \rho(u)|_{S_{i+\frac{1}{2}}} - S_{i-\frac{1}{2}} \rho(u)|_{S_{i-\frac{1}{2}}}\right] + \left[c_i u_i + \lambda[u_i - u_i^*]_+^{1/\kappa}\right] l_i = f_i l_i, \tag{2.3.1}$$

where $l_i = S_{i+1/2} - S_{i-1/2}, c_i = c(S_i, t), f_i = f(S_i, t), u_i^* = u^*(S_i, t)$, and u_i is the nodal approximation to $u(x_i, t)$ to be determined.

In Sect. 1.3, approximations to $\rho(u)$ at $S_{i+\frac{1}{2}}, i > 0$, and $S_{\frac{1}{2}}$ are constructed and given in (1.3.7) and (1.3.10) respectively. Using these, we have from (2.3.1)

$$-\frac{\partial u_i(t)}{\partial t} l_i + e_{i,i-1} u_{i-1}(t) + e_{i,i} u_i(t) + e_{i,i+1} u_{i+1}(t) + d_i(u_i(t), t) = f_i(t) l_i, \tag{2.3.2}$$

where $e_{i,i-1}, e_{i,i}$ and $e_{i,i+1}$ are defined in (1.3.16)–(1.3.19) and

$$d_i(u_i(t), t) = \lambda l_i [u_i(t) - u_i^*(t)]_+^{1/\kappa} \tag{2.3.3}$$

for $i = 1, 2, \ldots, N - 1$. These form an $N - 1$ nonlinear ODE system for $\boldsymbol{u}(t) := (u_1(t), \ldots, u_N(t))^\top$ with the homogeneous boundary condition $u_0(t) = 0 = u_N(t)$.

We now consider the time discretization of the linear ODE system (2.3.2). Let $E_i := (0, \ldots, 0, e_{i,i-1}(t), e_{i,i}(t), e_{i,i+1}(t), 0, \ldots, 0)$, where $e_{i,i-1}, e_{i,i}$ and $e_{i,i+1}$ are defined in (1.3.16)–(1.3.19) and those which are not defined are zeros. Using E_i, we rewrite (2.3.2) as the following matrix form:

$$-\frac{\partial u_i(t)}{\partial t} l_i + E_i(t)\boldsymbol{u}(t) + d_i(u_i(t), t) = f_i(t)l_i, \quad i = 1, 2, \ldots, N - 1. \quad (2.3.4)$$

To discretize the above ODE system, we choose a partition for $[0, T]$ with mesh nodes t_i ($i = 0, 1, \ldots, M$) satisfying $T = t_0 > t_1 > \cdots > t_M = 0$. Apply the two-level implicit time-stepping scheme with a splitting parameter $\theta \in [1/2, 1]$ to (2.3.4), we obtain, for $m = 0, 1, \ldots, M - 1$,

$$\frac{u_i^{m+1} - u_i^m}{-\Delta t_m} l_i + \theta \left[E_i^{m+1} \boldsymbol{u}^{m+1} + d_i^{m+1}(u_i^{m+1}) \right] + (1 - \theta) \left[E_i^m \boldsymbol{u}^m + d_i^m(u_i^m) \right]$$
$$= (\theta f_i^{m+1} + (1 - \theta) f_i^m) l_i,$$

where $\Delta t_m = t_{m+1} - t_m < 0$, $E_i^m = E_i(t_m)$, $f_i^m = f(x_i, t_m)$, $d_i^m(u_i^m) = d_i(u_i^m, t_m)$ and $\boldsymbol{u}^m = (u_1^m, u_2^m, \ldots, u_{N-1}^m)^\top$ denotes the approximation of $\boldsymbol{u}(t)$ at $t = t_m$. Let E^m be the $(N - 1) \times (N - 1)$ matrix given by $E^m = (E_1^m, E_2^m, \ldots, E_{N-1}^m)^\top$ and $D^m(\boldsymbol{u}^m) = (d_1^m(u_1^m), d_2^m(u_2^m), \ldots, d_{N-1}^m(u_{N-1}^m))^\top$. Then, the above system can be re-written as

$$(\theta E^{m+1} + G^m)\boldsymbol{u}^{m+1} + \theta D^{m+1}(\boldsymbol{u}^{m+1}) = \bar{f}^m + [G^m - (1 - \theta)E^m]\boldsymbol{u}^m - (1 - \theta)D^m(\boldsymbol{u}^m)$$
$$(2.3.5)$$

for $m = 0, 1, \ldots, M - 1$, where $G^m = \text{diag}(l_1/(-\Delta t_m), \ldots, l_{N-1}/(-\Delta t_m))$ is an $(N - 1) \times (N - 1)$ diagonal matrix and

$$\bar{f}^m = \theta(f_1^{m+1}l_1, \ldots, f_{N-1}^{m+1}l_{N-1})^\top + (1 - \theta)(f_1^m l_1, \ldots, f_{N-1}^m l_{N-1})^\top.$$

The boundary and terminal/payoff conditions for this system are

$$u_0^m = 0 = u_N^m, \quad m = 0, 1, \ldots, M, \quad \boldsymbol{u}^0 = (u_1^*(T), u_2^*(T), \ldots, u_{N-1}^*(T))^\top.$$

Remark 2.3.1 When $\theta = 1/2$, the time-stepping scheme becomes that of the Crank–Nicholson, and when $\theta = 1$, it is the backward Euler (or full implicit) scheme. Both are unconditionally stable. A stability and error analysis for this finite volume method with $D = 0$ in (2.3.5) is given in Sect. 1.3.3. For clarity, we will skip this discussion and refer the author to [2].

2.3.2 Solution of the Nonlinear System

Equation (2.3.5) is nonlinear in u^{m+1}. We now apply a Newton method to it. When $\kappa > 1$, from (2.3.3) we see that $d_i'(u_i^m) \to \infty$ as $u_i^m - u_i^* \to 0^+$. To overcome this difficulty, we smooth out $d_i(u_i^m)$ in the neighbourhood of $[u_i^m - u_i^*]_+ = 0$ by redefining d_i as

$$\frac{1}{\lambda l_i} d_i^m(u_i^m) = \begin{cases} (u_i^m - u_i^{*,m})^{1/\kappa}, & u_i^m - u_i^{*,m} \geq \varepsilon, \\ W([u_i^m - u_i^{*,m}]_+), & u_i^m - u_i^{*,m} < \varepsilon \end{cases} \tag{2.3.6}$$

for $k > 0$, where $1 \gg \varepsilon > 0$ is a transition parameter and $W(z)$ is a function which smooths out the original $d_i(z)$ around $z = 0$. We choose $W(z) = c_1 + c_2 z + \cdots + c_n z^{n-1} + c_{n+1} z^n$ for $n \geq 3$ and impose that $W(z)$ is such that $d_i(\cdot)$ is smooth. This requires that $W(z)$ satisfies $W(0) = W'(0) = 0$, $W(\varepsilon) = \varepsilon^{1/\kappa}$, and $W'(\varepsilon) = \frac{1}{k}\varepsilon^{1/\kappa-1}$. Thus, the function defined in (2.3.6) is globally smooth. Using these four conditions and setting $c_3 = \cdots = c_{n-1} = 0$, we can easily find that

$$c_1 = c_2 = 0, \quad c_n = \varepsilon^{1/\kappa-n+1}\left(n - \frac{1}{\kappa}\right), \quad c_{n+1} = \varepsilon^{1/\kappa-n}\left(\frac{1}{\kappa} - n + 1\right).$$

Substituting these into (2.3.6) gives

$$\frac{1}{\lambda l_i} d_i^m(u_i^m) = \begin{cases} (u_i^m - u_i^{*,m})^{1/\kappa}, & u_i^m - u_i^{*,m} \geq \varepsilon, \\ \varepsilon^{1/\kappa-n+1}\left(n - \frac{1}{\kappa}\right)[u_i^m - u_i^{*,m}]_+^{n-1} \\ + \varepsilon^{1/\kappa-n}\left(\frac{1}{\kappa} - n + 1\right)[u_i^m - u_i^{*,m}]_+^n, & u_i^m - u_i^{*,m} < \varepsilon. \end{cases} \tag{2.3.7}$$

We comment that the intuition of this choice of d_i^m is as follows. When $u_i^m - u_i^{*,m} \geq \varepsilon$, a given tolerance, $(u_i^m - u_i^{*,m})^{1/\kappa}$ offers a convergence rate of order $\kappa/2$ by the results from the previous section. When $[u_i^m - u_i^{*,m}] < \varepsilon$, we choose $d_i^m(u_i^m) = W([u_i^m - u_i^{*,m}]_+)$ to slow down the convergence. For the function $W(z)$ used to smooth the penalty term, we have the following theorem.

Theorem 2.3.1 *The function $W(z)$ is strictly increasing on $[0, \varepsilon]$ when $\kappa \geq 1/n$ and $n \geq 3$.*

Proof When $z = 0$, $W(0) = 0$. We now show that $W'(z) > 0$ for $z \in (0, \varepsilon)$. Differentiating $W(z)$ gives

$$W'(z) = (n-1)\varepsilon^{1/\kappa-n+1}\left(n - \frac{1}{\kappa}\right)z^{n-2} + n\varepsilon^{1/\kappa-n}\left(\frac{1}{\kappa} - n + 1\right)z^{n-1}$$

$$= z^{n-2}\varepsilon^{1/\kappa-n+1}\left[(n-1)\left(n - \frac{1}{\kappa}\right) + n\left(\frac{1}{\kappa} - n + 1\right)\frac{z}{\varepsilon}\right]$$

$$= z^{n-2}\varepsilon^{1/\kappa-n+1}G(z),$$

where $G(z) := (n-1)\left(n - \frac{1}{\kappa}\right) + n\left(\frac{1}{\kappa} - n + 1\right)\frac{z}{\varepsilon}$.

We now show $G(z) > 0$ on $(0, \varepsilon]$ when $\kappa \geq 1/n$. In fact, $G(0) = (n - 1)\left(n - \frac{1}{\kappa}\right)$ ≥ 0 and $G(\varepsilon) = 1/\kappa > 0$. Since $G(z)$ is linear, we have $G(z) > 0$ on $(0, \varepsilon]$. Therefore, we have $W'(z) > 0$ on $(0, \varepsilon)$, and so $W(z)$ is strictly increasing on $[0, \varepsilon]$. $\qquad \square$

Using this theorem, we have the following corollary.

Corollary 2.3.1 *The nonlinear function $d_i^m(u_i^k)$ defined in (2.3.7) is smooth and increasing on $(-\infty, \infty)$ when $k \geq 1/n$ and $n \geq 3$.*

Proof By Theorem 2.3.1, $d_i^m(u_i^m)$ is increasing in the region $(u_i^*, u_i^* + \varepsilon)$. Also, from (2.3.7) we see that $d(u_i^m) = (u_i^m - u_i^{*,m})^{1/k}$ if $u_i^k \geq u_i^* + \varepsilon$ and $d(u_i^m) = 0$ if $u_i^m \leq u_i^*$. Both of these are increasing. The smoothness of the function is obvious because of the choice of W. This completes the proof. $\qquad \square$

We use the following Newton's method to (2.3.5) with d_i^m defined in (2.3.7).

Algorithm Newton

1. Choose penalty and smoothing parameters $\lambda > 1, \kappa \geq 1$ and $0 < \varepsilon \ll 1$, tolerance $0 < \delta \ll 1$, and damping parameter $\gamma \in (0, 1]$. Calculate terminal condition u^0 and let $m = 0$.
2. Let $w^0 = u^m$ and $l = 1$.
3. Solve the following system for w^l.

$$[\theta E^{m+1} + G^m + \theta J_{D^m}(w^{l-1})]\Delta^l = \bar{f}^m + [G^m - (1 - \theta)E^m]u^m - (1 - \theta)D^m(u^m)$$
$$- (\theta E^{m+1} + G^m)w^{l-1} - \theta D^m(w^{l-1}), \qquad (2.3.8)$$
$$w^l = w^{l-1} + \gamma \Delta^l,$$

where $J_{D^m}(w)$ denotes the Jacobian of the column vector $D^m(w)$.
4. If $\|w^l - w^{l-1}\|_2 \leq \delta$, then go to Step 5, where $\|\cdot\|_2$ denotes the Euclidean norm. Otherwise, let $l := l + 1$ and go to Step 3.
5. Let $u^{m+1} = w^l$. If $m = M - 1$, then stop and the problem is solved. Otherwise, let $m := m + 1$ and go to Step 2.

For the system matrix of (2.3.8), we have the following theorem.

Theorem 2.3.2 *The system matrix $\theta E^{m+1} + G^m + \theta J_{D^m}(w^{l-1})$ of (2.3.8) is an M-matrix when $|\Delta t| := \max_m |\Delta t_m|$ is sufficiently small.*

Proof We have already shown in Theorem 1.3.1 that $\theta E^{m+1} + G^m$ is an M-matrix. Thus we only need to show that the diagonal entries of J_{D^m} are all non-negative.

From the definition of D^m in (2.3.5), we see that, for any w, its Jacobian is

$$J_{D^m}(w) = \text{diag}((d_1^m)'(w_1), (d_2^m)'(w_2), \ldots, (d_{N-1}^m)'(w_{N-1})).$$

By Corollary 2.3.1, we have $d_i'(w_i) \geq 0$ for $i = 1, 2, \ldots, N - 1$, and thus the diagonal entries of J_{D^m} are non-negative. Combining this with Theorem 1.3.1 yields $\theta E^{m+1} + G^m + \theta J_{D^m}(w^{l-1})$ is an M-matrix. $\qquad \square$

2.4 Numerical Experiments

We now use the methods developed in the previous sections to solve some model problems. As in Sect. 1.4, for all the tests given below, we choose $S_{\max} = 100$ and $T = 1$. The uniform mesh with 51×51 mesh nodes is used for solving the test unless mentioned otherwise.

Example 2.1 The American put option with $K = 50$. The market parameters are $\sigma = 0.4, r = 0.03, d = 0.02$.

To solve this problem, we choose $\lambda = 10^4$ and $\kappa = 2$ in (2.2.1). The computed option value V, its Greeks Δ and $\Gamma := \frac{\partial^2 V}{\partial S^2}$, and $V - V^*$ are plotted in Fig. 2.1. From Δ and Γ in Fig. 2.1 we see that an interior curve can clearly be seen, while the Δ of its European counterpart depicted in Fig. 1.2 does not contain such a curve. To further demonstrate the optimal exercise curve, we have also solved this problem on the 301×301 uniform mesh and use this numerical solution to estimate the optimal exercise curve, which is depicted in the V plot in Fig. 2.1.

We now show the influence of κ on the quality of solutions. We choose $\lambda = 5$ and solve the problem for $\kappa = 1, 3, 5$. The computed V and Δ for each κ are plotted in

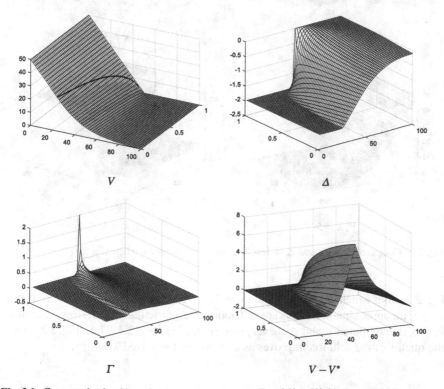

Fig. 2.1 Computed value V, optimal exercise curve, Δ, Γ and $V - V^*$ for Example 2.1

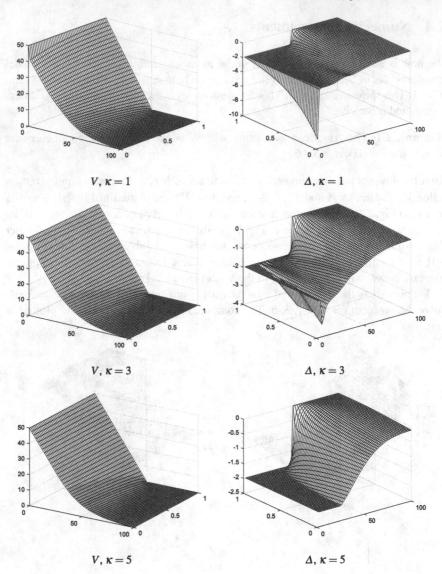

Fig. 2.2 Computed value V and Δ of Test 2 for various values of κ when $\lambda = 5$

Fig. 2.2, from which we see that when $\kappa = 1$, there is a large error near $S = 0$. When $\kappa = 3$, the approximation errors are still noticeable, particularly in Δ. However, when $\kappa = 5$, the computed V and Δ are very close to those in Fig. 2.1. Thus, Fig. 2.2 shows the quality of the solution improves as κ increases for a fixed λ.

References

1. Achdou Y, Pironneau O (2005) Computational methods for option pricing. SIAM, New York
2. Angermann L, Wang S (2007) Convergence of a fitted finite volume method for the penalized Black-Scholes equations governing European and American option pricing. Numer Math 106:1–40
3. Bensoussan A, Lions JL (1982) Applications of variational inequalities in stochastic control. North-Holland, Amsterdam
4. Benth FE, Karlsen KH, Reikvam K (2004) A semilinear Black and Scholes partial differential equation for valuing American options: approximate solutions and convergence. Interfaces Free Bound 6:379–404
5. Forsyth PA, Vetzal KR (2002) Quadratic convergence for valuing American options using a penalty method. SIAM J Sci Comput 23:2095–2122
6. Glowinski R (1984) Numerical methods for nonlinear variational problems. Springer, New York
7. Han H, Wu X (2003) A fast numerical method for the Black-Scholes equation of American options. SIAM J Numer Anal 41:2081–2095
8. Haslinger J, Miettinen M (1999) Finite Element Method for Hemivariational Inequalities. Kluwer Academic Publisher, Dordrecht
9. Kinderlehrer D, Stampacchia G (1980) An introduction to variational inequalities and their applications. Academic, New York
10. Nielsen BF, Skavhaug O, Tveito A (2001) Penalty and front-fixing methods for the numerical solution of American option problems. J Comp Fin 5:69–97
11. Wang G, Wang S (2006) On stability and convergence of a finite difference approximation to a parabolic variational inequality arising from American option valuation. Stoch Anal Appl 24:1185–1204
12. Wang G, Wang S (2010) Convergence of a finite element approximation to a degenerate parabolic variational inequality with non-smooth data arising from American option valuation. Optim Methods Softw 25:699–723
13. Wang S, Yang XQ, Teo KL (2006) Power penalty method for a linear complementarity problem arising from American option valuation. J Optim Theory Appl 129:227–254

Chapter 3
Options on One Asset with Stochastic Volatility

Abstract In this chapter we develop numerical methods for pricing European and American options whose underlying asset price and volatility follow two separate geometric Brownian motions. These methods include a fitted Finite Volume Method (FVM) for the discretization of the resulting 2D Black–Scholes equation and a power penalty approach to the differential Linear Complementarity Problem involving the 2D differential operator of Black–Scholes type. A mathematical analysis will be presented for the convergence of the FVM and power penalty approach. These methods can also be used for pricing options on two assets such as a basket option.

Keywords Option pricing under two-factor models · Stochastic volatility · 2-dimensional Black–Scholes equation · Linear complementarity problem · Finite volume methods, penalty method

3.1 The 2-Dimensional PDE Model for Pricing European Options with Stochastic Volatility

There are different stochastic volatility models in option pricing, such as Heston's model [6] and the Constant Elasticity of Variance (CEV) Model [1]. In what follows, we shall consider the latter which is widely used by researchers (e.g., [5]). We will first derive the PDE governing the pricing problem. We will then formulate the 'strong' problem as an variational one and show that the latter is uniquely solvable.

3.1.1 The Pricing Problem

Assume a stock price S and its instantaneous volatility $\sigma =: \sqrt{v}$ satisfy the following stochastic equations.

$$dS = \mu S dt + \sqrt{v} S dW_1 \text{ and } dv = \mu_v v dt + \sigma_v v dW_2, \tag{3.1.1}$$

© The Author(s), under exclusive license to Springer Nature Singapore Pte Ltd. 2020 55
S. Wang, *The Fitted Finite Volume and Power Penalty Methods*
for Option Pricing, SpringerBriefs in Mathematical Methods,
https://doi.org/10.1007/978-981-15-9558-5_3

where μ is the drift coefficient for S, μ_v and σ_v are the constant drift coefficient and volatility of v, and W_1 and W_2 are two Wiener processes with correlation ρ.

It is shown in [2] that a security F with a price depending on assets $S_i, i = 1, 2, \ldots, N$ satisfies the following PDE (also see [5]):

$$\frac{\partial F}{\partial t} + \frac{1}{2} \sum_{i,j=1}^{N} \rho_{ij} \sigma_i \sigma_j \frac{\partial^2 F}{\partial S_i \partial S_j} - rF = \sum_{i=1}^{N} S_i \frac{\partial F}{\partial S_i} [-\mu_i + \beta_i (\mu^* - r)], \quad (3.1.2)$$

where σ_i is the instantaneous standard deviation of S_i, ρ_{ij} is the instantaneous correlation between S_i and S_j, μ_i is the expected return rate of S_i, $\{\beta_i\}$ the vector of instantaneous sensitivity (or betas) of the expected asset returns (dS/S) to the expected market return, and μ^* is the instantaneous expected return on the market portfolio. When S_i is traded, it should satisfy the continuous time $(N + 1)$-factor Capital Asset Pricing Model (CAPM). Therefore, its return rate should satisfy $\mu_i = r + \beta_i (\mu^* - r)$ [4, Page 795]. Thus, the ith term of the sum on the RHS of (3.1.2) becomes $-rS_i \frac{\partial F}{\partial S_i}$.

In our Problem (3.1.1), there are two assets S and v. Thus, if V is a European option with strike price K and maturity T whose price depends on S and v, using (3.1.2) and the above analysis we see that V satisfies the following equation.

$$\frac{\partial V}{\partial t} + \frac{1}{2} \left[vS^2 \frac{\partial^2 V}{\partial S^2} + 2\rho \sigma_v S v^{3/2} \frac{\partial^2 V}{\partial S \partial v} + \sigma_v^2 v \frac{\partial^2 V}{\partial v^2} \right] - rV$$
$$= -rS \frac{\partial V}{\partial S} - [\mu_v - \beta_v (\mu^* - r)] v \frac{\partial V}{\partial v}, \quad (3.1.3)$$

where β_v denotes the sensitivity (beta) of the expected return of v. As in [5], we assume $\beta_v (\mu^* - r) = 0$, i.e., the volatility is uncorrelated with aggregated consumption, or the volatility has zero systematic risk. Thus, (3.1.3) becomes the following 2-dimensional Black–Scholes equation:

$$-\frac{\partial V}{\partial t} - \frac{1}{2} \left[vS^2 \frac{\partial^2 V}{\partial S^2} + 2\rho \sigma_v S v^{3/2} \frac{\partial^2 V}{\partial S \partial v} + \sigma_v^2 v \frac{\partial^2 V}{\partial v^2} \right] - rS \frac{\partial V}{\partial S} - \mu_v v \frac{\partial V}{\partial v} + rV = 0$$
$$(3.1.4)$$

for $S > 0$, $v > 0$ and $t \in [0, T)$. It is easy to show [7] that (3.1.4) can be written as

$$-\frac{\partial V}{\partial t} - \nabla \cdot (A \nabla V + bV) + cV = 0, \quad (3.1.5)$$

where $A = \begin{pmatrix} a_{11} & a_{12} \\ a_{21} & a_{22} \end{pmatrix}$, $b = (b_1, b_2)^\top$ and c is scalar with

$$a_{11} = \frac{1}{2} S^2 v, \quad a_{22} = \frac{1}{2} \sigma_v^2 v^2, \quad a_{12} = a_{21} = \frac{1}{2} \rho \sigma_v S v^{3/2}, \quad (3.1.6)$$

$$b_1 = rS - \frac{3}{4} \rho \sigma_v S v - Sv, \quad b_2 = \mu_v v - \frac{1}{2} \rho \sigma_v v^{3/2} - \sigma_v^2 v \quad (3.1.7)$$

$$c = 2r - \frac{3}{2}\rho\sigma_v v^{1/2} - v + \mu_v - \sigma_v^2. \tag{3.1.8}$$

To solve (3.1.4) numerically, we need to truncate its solution domain and define boundary and payoff conditions for it. Let $S_{\max}$, $v_{\min}$ and v_{max} be positive integers satisfying $S_{\max} > K$ and $0 < v_{\min} < v_{\max}$. Let $\kappa(V) := A\nabla V + bV$. Following [7], we rewrite (3.1.5) as the following problem:

$$-\frac{\partial V}{\partial t} - \nabla \cdot \kappa(V) + cV = 0, \tag{3.1.9}$$

$$V(0, v, t) = g_1(t), \ V(S_{\max}, v, t) = g_2(t), \ V(S, v, T) = g_3(S), \tag{3.1.10}$$

$$V(S, v_{\min}, t) = G_1(S, t), \ V(S, v_{\max}, t) = G_2(S, t) \tag{3.1.11}$$

for $(S, v, t) \in \Omega :=\in (0, S_{\max}) \times (v_{\min}, v_{\max}) \times [0, T)$, where g_i, $i = 1, 2, 3$ in (3.1.10) are the same as defined in (1.2.6)–(1.2.10). The boundary conditions G_1 and G_2 in (3.1.11) can not normally be found explicitly, but we can calculate them numerically by solving (1.2.3)–(1.2.4) in Sect. 1.2.2 at $\sigma^2 = v_{\min}$ and $v_{\max}$.

3.1.2 The Variational Problem and Its Solvability

The discussions below follow those in [8]. Without loss of generality, we assume that g_1, g_2, G_1 and G_2 in (3.1.10)–(3.1.11) are all zero. The case with non-zero boundary conditions can be transformed into one with homogeneous boundary conditions by subtracting a known smooth function V_0 satisfying the non-homogeneous boundary conditions (e.g., the solution of $-\nabla^2 V_0 = 0$ satisfying the boundary conditions in (3.1.10)–(3.1.11)). This transformation creates a non-zero RHS function. We introduce the transformation $u = e^{\beta t}(V - V_0)$ for any constant $\beta > 0$. Under this transformation, (3.1.9)–(3.1.11) can be written as the following problem:

$$\mathscr{L}_{2D}u := -\frac{\partial u}{\partial t} - \nabla \cdot \kappa(u) + \hat{c}u = f, \tag{3.1.12}$$

$$u(S, v, t) = 0, \ (S, v, t) \in \partial\Omega \times [0, T), \ u(S, v, T) = \hat{g}_3(S, v), \ (S, v) \in \Omega, \tag{3.1.13}$$

where $\hat{c} = \beta + c$ and $\partial\Omega$ denotes the boundary of Ω.

Before further discussion, we first introduce some notation to be used in the analysis. Let $L^2(\Omega)$ be the space of all square-integral functions on Ω. We define an inner product and norm on $L^2(\Omega)$ by $(u, v) := \int_\Omega u^\top v d\Omega$ and $\|v\|_0 = \sqrt{(v, v)}$ respectively for $u, v \in L^2(\Omega)$.

We now introduce a weighted inner product on $(L^2(\Omega))^2$ defined by $(u, v)_w := \int_\Omega (S^2 u_1 v_1 + v^2 u_2 v_2) d\Omega$ for any $u = (u_1, u_2)^\top, v = (v_1, v_2)^\top \in (L^2(\Omega))^2$. Let

$\|v\|_{0,w} := \sqrt{(v, v)_w}$ be the norm generated by $(\cdot, \cdot)_w$. We define a space of all weighted square-integrable functions in $(L^2(\Omega))^2$ by $\mathbf{L}_w^2(\Omega) := \{v \in (L^2(\Omega))^2 : \|\nabla v\|_{0,w} < \infty\}$.

Using a standard argument it is easy to show that the pair $(\mathbf{L}_w^2(\Omega), (\cdot, \cdot)_w)$ is a Hilbert space (e.g., [9]). Based on this space, we define a weighted Sobolev space $H_w^1(\Omega) = \{v \in L^2(\Omega) : \nabla v \in \mathbf{L}_w^2(\Omega)\}$ with the energy norm defined by $\|v\|_{1,w} = (|v|_{1,w}^2 + \|v\|_0^2)^{1/2}$, where $|v|_{1,w} = \|\nabla v\|_{0,w}$. is a semi-norm. Let $\partial \Omega_D = \{(S, v) \in \partial\Omega : S \neq 0\}$ denote the boundary segments of Ω excluding the part where $S = 0$. Let $H_{0,w}^1(\Omega) = \{v \in H_w^1(\Omega) : v|_{\partial\Omega_D} = 0\}$.

Multiplying (3.1.12) by $v \in H_{0,w}^1(\Omega)$ and applying integration by parts yield

$$-\left(\frac{\partial u}{\partial t}, v\right) - \int_{\partial\Omega} v\kappa(u) \cdot n\, ds + (\kappa(u), \nabla v) + (\hat{c}u, v) = (f, v),$$

where n denotes the unit vector outward-normal to $\partial\Omega$. Since $v|_{\partial\Omega_D} = 0$, from the definition of $\kappa(u)$ and (3.1.6)–(3.1.7) we have

$$\int_{\partial\Omega} v\kappa(u) \cdot n\, ds = -\int_{\partial\Omega \setminus \partial\Omega_D} \left(a_{11}\frac{\partial u}{\partial S} + a_{12}\frac{\partial u}{\partial v} + b_1 u\right) v\, ds = 0 \qquad (3.1.14)$$

since a_{11}, a_{12} and b_1 are all equal to zero at $S = 0$. Therefore, if we define the following bi-linear form on $H_w^1(\Omega)$

$$B(u, v; t) = (A\nabla u + bu, \nabla v) + (\hat{c}u, v), \qquad (3.1.15)$$

we then propose the following variational problem.

Problem 3.1.1 Find $u(t) \in H_{0,w}^1(\Omega)$, satisfying the final condition in (3.1.13), such that for all $v \in H_{0,w}^1(\Omega)$,

$$\left(-\frac{\partial u(t)}{\partial t}, v\right) + B(u(t), v; t) = (f, v) \quad \text{a.e. in } (0, T). \qquad (3.1.16)$$

From the above analysis we see that Problem 3.1.1 is the variational problem corresponding to (3.1.12)–(3.1.13). Note that we do not need the boundary condition that $V(0, v, t) = 0$ due to the degeneracy of $\mathcal{L}_{2D}$ at $S = 0$. In this case, a solution to Problem 3.1.1 cannot take a 'trace' (boundary condition) at $S = 0$. However, in computation, we choose a particular solution with a given value at $S = 0$.

The unique solvability of Problem 3.1.1 is given in the following theorem.

Theorem 3.1.1 *Problem 3.1.1 has a unique solution.*

Proof We show B is coercive and continuous. For any $\phi \in H_{0,w}^1(\Omega)$, using integration by parts and (3.1.14), we have

$$(b\phi, \nabla\phi) = \int_{\partial\Omega} \phi^2 b \cdot n\, ds - \int_{\Omega} \phi \nabla \cdot (b\phi)\, d\Omega = -\int_{\Omega} \phi b \cdot \nabla\phi\, d\Omega - \int_{\Omega} \phi^2 \nabla \cdot b\, d\Omega.$$

Thus, $(b\phi, \nabla\phi) = -\frac{1}{2}(\phi\nabla \cdot b, \phi)$, and so from (3.1.15) and (3.1.6)–(3.1.8), we have

$$
B(\phi, \phi; t) = (A\nabla\phi, \nabla\phi) + \left(\left(\hat{c} - \frac{1}{2}\nabla \cdot \underline{b}\right)\phi, \phi\right)
$$

$$
= \frac{1}{2}\int_\Omega \left[vS^2\phi_S^2 + 2\rho v^{\frac{1}{2}}\sigma_v vS\phi_S\phi_v + \sigma_v^2 v^2\phi_v^2\right]d\Omega + \left(\left(\hat{c} - \frac{1}{2}\nabla \cdot b\right)\phi, \phi\right)
$$

$$
= \frac{1}{2}\int_\Omega \left[(1-\rho)vS^2\phi_S^2 + \rho(v^{\frac{1}{2}}S\phi_S + \sigma_v v\phi_v)^2 + (1-\rho)\sigma_v^2 v^2 v\phi_v^2\right]d\Omega
$$

$$
+ ((\hat{c} - \frac{1}{2}\nabla \cdot b)\phi, \phi) \geq \left(\left(\beta + \frac{3}{2}r - \frac{1}{2}\left(\frac{3}{2}\rho\sigma_v v^{\frac{1}{2}} + y - \mu_v + \sigma + v^2\right)\right)\phi, \phi\right)
$$

$$
+ C\int_\Omega \left[S^2\phi_S^2 + v^2\phi_v^2\right]d\Omega \geq C\|\phi\|_{1,w}^2 \tag{3.1.17}
$$

when β is chosen to be sufficiently large, where C denotes a generic positive constant, independent of v. Therefore, B is coercive.

To show $B(\phi, \psi; t)$ is continuous, we first note that, for $\phi, \psi \in H_{0,w}^1(\Omega)$, we have

$$
|B(\phi, \psi; t)| \leq |(A\nabla\phi, \nabla\psi)| + |(b\phi, \nabla\psi)| + |(\hat{c}\phi, \psi)|. \tag{3.1.18}
$$

For $|(A\nabla\phi, \nabla\psi)|$ in (3.1.18), using the triangular inequality, we have

$$
|(A\nabla\phi, \nabla\psi)| \leq \frac{1}{2}\underbrace{\left|\int_\Omega (vS^2\phi_S\psi_S + \sigma_v^2 v^2\phi_v\psi_v)d\Omega\right|}_{I_1}
$$

$$
+ \frac{1}{2}\underbrace{\left|\int_\Omega \rho\sigma_v Sv^{\frac{3}{2}}(\phi_S\psi_v + \phi_v\psi_S)d\Omega\right|}_{I_2} \tag{3.1.19}
$$

We now estimate I_1 and I_2. Using Cauchy–Schwarz inequality, we obtain

$$
I_1 \leq \left|\int_\Omega vS^2\phi_S\psi_S d\Omega\right| + \left|\int_\Omega \sigma_v^2 v^2\phi_v\psi_v d\Omega\right| = \left|(v^{\frac{1}{2}}S\phi_S, v^{\frac{1}{2}}S\psi_S)\right| + |(\sigma_v v\phi_v, \sigma_v v\psi_v)|
$$

$$
\leq (v^{\frac{1}{2}}S\phi_S, v^{\frac{1}{2}}S\phi_S)^{\frac{1}{2}} \cdot (v^{\frac{1}{2}}S\psi_S, v^{\frac{1}{2}}S\psi_S)^{\frac{1}{2}} + (\sigma_v v\phi_v, \sigma_v v\phi_v)^{\frac{1}{2}} \cdot (\sigma_v v\psi_v, \sigma_v v\psi_v)^{\frac{1}{2}}
$$

$$
= \left[\int_\Omega vS^2\phi_S^2 d\Omega\right]^{\frac{1}{2}} \cdot \left[\int_\Omega vS^2\psi_S^2 d\Omega\right]^{\frac{1}{2}} + \left[\int_\Omega \sigma_v^2 v^2\phi_v^2 d\Omega\right]^{\frac{1}{2}} \cdot \left[\int_\Omega \sigma_v^2 v^2\psi_v^2 d\Omega\right]^{\frac{1}{2}}
$$

$$
\leq \left[\int_\Omega (vS^2\phi_S^2 + \sigma_v^2 v^2\phi_v^2)d\Omega\right]^{\frac{1}{2}} \cdot \left[\int_\Omega (vS^2\psi_S^2 + \sigma_v^2 v^2\psi_y^2)d\Omega\right]^{\frac{1}{2}} \leq M|\phi|_{1,w}|\psi|_{1,w},
$$

where M denotes a generic positive constant, independent of ϕ and ψ.

Similarly, we have

$$
\begin{aligned}
I_2 &\le \left| \int_\Omega \rho\sigma_v S v^{\frac{3}{2}} \phi_S \psi_v d\Omega \right| + \left| \int_\Omega \rho\sigma_v S v^{\frac{3}{2}} \phi_v \psi_S d\Omega \right| \\
&= \left| (\rho v^{\frac{1}{2}} S \phi_S, \sigma_v v \psi_v) \right| + \left| (\rho\sigma_v v \phi_v, v^{\frac{1}{2}} S \psi_S) \right| \\
&\le M \left\{ \left[\int_\Omega v S^2 \phi_S^2 d\Omega \right]^{\frac{1}{2}} \cdot \left[\int_\Omega v^2 \psi_v^2 d\Omega \right]^{\frac{1}{2}} + \left[\int_\Omega v^2 \phi_v^2 d\Omega \right]^{\frac{1}{2}} \cdot \left[\int_\Omega v S^2 \psi_S^2 d\Omega \right]^{\frac{1}{2}} \right\} \\
&\le M \left[\int_\Omega (v S^2 \phi_S^2 + v^2 \phi_v^2) d\Omega \right]^{\frac{1}{2}} \cdot \left[\int_\Omega (v S^2 \psi_S^2 + v^2 \psi_v^2) d\Omega \right]^{\frac{1}{2}} \le M |\phi|_{1,w} |\psi|_{1,w}.
\end{aligned}
$$

Thus, using the above estimates for I_1 and I_2, we obtain from (3.1.19)

$$
|(A\nabla\phi, \nabla\psi)| \le M |\phi|_{1,w} |\psi|_{1,w}. \tag{3.1.20}
$$

For $|(\underline{b}v, \nabla w)| = |\int_\Omega b\phi \cdot \nabla\psi d\Omega|$ in (3.1.18), using integration by parts gives

$$
\begin{aligned}
|(b\phi, \nabla\psi)| &= \left| \int_{\partial\Omega} \phi\psi b \cdot n ds - \int_\Omega \psi \nabla \cdot (b\phi) d\Omega \right| = \left| -\int_\Omega \psi b \cdot \nabla\phi d\Omega - \int_\Omega \phi\psi \nabla \cdot b d\Omega \right| \\
&\le \left| \int_\Omega \psi b \cdot \nabla\phi d\Omega \right| + \left| \int_\Omega \phi\psi \nabla \cdot b d\Omega \right| \le \left| \int_\Omega \psi b \cdot \nabla\phi d\Omega \right| + M \|v\|_0 \|w\|_0.
\end{aligned}
$$

Furthermore, from (3.1.15) and by Cauchy–Schwarz inequality, we obtain

$$
\begin{aligned}
\left| \int_\Omega \psi b \cdot \nabla\phi d\Omega \right| &= \left| \int_\Omega \psi \left[\left(r - \frac{3}{4} \rho v^{\frac{1}{2}} \sigma_v - v \right) S v_S + \left(\mu_v - \frac{1}{2} \rho v^{\frac{1}{2}} \sigma_v - \sigma_v^2 \right) v \phi_v \right] d\Omega \right| \\
&\le M \left| \int_\Omega \psi(S\phi_S + v\phi_v) d\Omega \right| \le M \left(\int_\Omega \psi^2 d\Omega \right)^{\frac{1}{2}} \cdot \left(\int_\Omega (S\phi_S + v\phi_v)^2 d\Omega \right)^{\frac{1}{2}} \\
&\le M \|w\|_0 |v|_{1,w}.
\end{aligned}
$$

Therefore, combining the above two estimates, we have,

$$
|(b\phi, \nabla\psi)| \le M \left(\|\psi\|_0 |\phi|_{1,w} + \|v\|_0 \|w\|_0 \right). \tag{3.1.21}
$$

It is trivial to show $|(\hat{c}\phi, \psi)| \le M \|\phi\|_0 \|\psi\|_0$. Thus, Combining (3.1.18), (3.1.20) and (3.1.21) estimate, we have

$$
|B(\phi, \psi; t)| \le M \left(|\phi|_{1,w} |\psi|_{1,\mathbf{w}} + \|\psi\|_0 |\phi|_{1,w} + \|\phi\|_0 \|\psi\|_0 \right) \le M \|\phi\|_{1,\mathbf{w}} \|w\|_{1,w}.
$$

Therefore, $\mathbf{B}(\phi, \psi; t)$ is Lipschitz continuous in ϕ and ψ.

Since $B(\cdot, \cdot; t)$ is coercive and continuous, from [3, Theorem 1.33], we see that Problem 3.1.1 has a unique solution, completing the proof. □

3.2 The Fitted FVM for (3.1.9)–(3.1.11)

We now present the FVM for (3.1.5) proposed in [7]. In what follows, we assume that r and μ_ν are also functions of t and σ_ν is a function of S and t.

For given positive integers N_S and N_ν, we define a mesh for Ω with mesh nodes

$$0 = S_0 < S_1 < \cdots < S_{N_S} = S_{max}, \quad \nu_{min} = \nu_0 < \nu_1 < \cdots < \nu_{N_\nu} = \nu_{max}. \quad (3.2.1)$$

Dual to this mesh, we define a secondary mesh for Ω with nodes $(S_{i+\frac{1}{2}}, \nu_{j+\frac{1}{2}})$ for $i = -1, 0, 1, \ldots, N_S$ and $j = -1, 0, 1, \ldots, N_\nu$, where $S_{-\frac{1}{2}} = 0$, $S_{N_S+\frac{1}{2}} = S_{max}$, $\nu_{-\frac{1}{2}} = \nu_{min}$, $\nu_{N_\nu+\frac{1}{2}} = \nu_{max}$, $S_{i+\frac{1}{2}} = \frac{1}{2}(S_i + S_{i+1})$ for $i = 0, 1, \ldots, N_S - 1$ and $\nu_{j+\frac{1}{2}} = \frac{1}{2}(\nu_j + \nu_{j+1})$ for $j = 0, 1, \ldots, N_\nu - 1$. For each $i = 0, 1, \ldots, N_S$ and $j = 0, 1, \ldots, N_\nu$, we put $h_{S_i} = S_{i+1/2} - S_{i-1/2}$ and $h_{\nu_j} = \nu_{j+1/2} - \nu_{j-1/2}$. A typical local stencil of the meshes is depicted in Fig. 3.1 in which $\Omega_{ij} := (S_{i-\frac{1}{2}}, S_{i+\frac{1}{2}}) \times (\nu_{j-\frac{1}{2}}, \nu_{j+\frac{1}{2}})$.

Integrating (3.1.9) over Ω_{ij} and using integration by parts, we have

$$-\int_{\Omega_{ij}} \frac{\partial V}{\partial t} d\Omega - \int_{\partial\Omega_{ij}} \kappa(V) \cdot n ds + \int_{\Omega_{ij}} cV d\Omega = 0$$

for $i = 1, 2, \ldots, N_S - 1$ and $j = 1, 2, \ldots, N_\nu - 1$, where $\partial\Omega_{ij}$ denotes the boundary of Ω_{ij} ans n denotes the unit vector outward-normal to $\partial\Omega_{ij}$. Applying the one-point quadrature rule to the first and third terms, we obtain from the above

$$-\frac{\partial V_{i,j}}{\partial t}|\Omega_{ij}| - \int_{\partial\Omega_{ij}} \kappa(V) \cdot n ds + c_{i,j} V_{i,j}|\Omega_{ij}| = 0, \quad (3.2.2)$$

where $|\cdot|$ denote the measure or absolute value, depending on the context, $c_{i,j} = c(S_i, \nu_j, t)$, and $V_{i,j}$ denotes the nodal approximation to $V(S_i, \nu_j, t)$.

We now consider the approximation of the 2nd term in (3.2.2). Note that $\partial\Omega_{ij}$ is the boundary of the rectangle oriented counter-clockwise as depicted Fig. 3.1 by dashed lines. Thus, from the definition of κ we have

Fig. 3.1 A local structure of the meshes

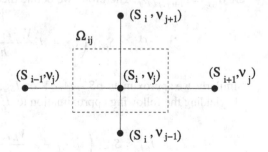

$$\int_{\partial \Omega_{ij}} \kappa(V) \cdot nds = \left(\int_{(S_{i+\frac{1}{2}}, \nu_{j-\frac{1}{2}})}^{(S_{i+\frac{1}{2}}, \nu_{j+\frac{1}{2}})} - \int_{(S_{i-\frac{1}{2}}, \nu_{j-\frac{1}{2}})}^{(S_{i-\frac{1}{2}}, \nu_{j+\frac{1}{2}})} \right) \left(a_{11} \frac{\partial V}{\partial S} + a_{12} \frac{\partial V}{\partial \nu} + b_1 V \right) d\nu$$

$$+ \left(\int_{(S_{i-\frac{1}{2}}, \nu_{j-\frac{1}{2}})}^{(S_{i+\frac{1}{2}}, \nu_{j-\frac{1}{2}})} - \int_{(S_{i-\frac{1}{2}}, \nu_{j+\frac{1}{2}})}^{(S_{i+\frac{1}{2}}, \nu_{j+\frac{1}{2}})} \right) \left(a_{21} \frac{\partial V}{\partial S} + a_{22} \frac{\partial V}{\partial \nu} + b_2 V \right) dS$$

$$=: J_{i,j}^1 - J_{i,j}^2 + J_{i,j}^3 - J_{i,j}^4. \tag{3.2.3}$$

To approximate $J_{i,j}^1$ and $J_{i,j}^2$ on the RHS of (3.2.3), we first develop approximations to their integrand at the mid-points of $I_{S_{i-1}}$ and I_{S_i}, where $I_{S_k} := (S_k, S_{k+1})$. Let us consider a constant approximation to this integrand on i_{S_k}, $k = i, i - 1$, in the following two cases.

Case 1. Approximation of J_{ij}^1 and J_{ij}^2 when $i > 1$.

Using (3.1.6)–(3.1.7), we rewrite the integrand as

$$a_{11} \frac{\partial V}{\partial S} + a_{12} \frac{\partial V}{\partial \nu} + b_1 V = S \left(p(\nu) S \frac{\partial V}{\partial S} + q(\nu) V + d(\nu) \frac{\partial V}{\partial \nu} \right),$$

where $p = \frac{1}{2}\nu, q = r - \frac{3}{4}\rho\nu^{\frac{1}{2}}\sigma_\nu - \nu$ and $d = \frac{1}{2}\rho\sigma_\nu\nu^{\frac{3}{2}}$. Following the discussion in Sect. 1.3.1, we approximate the term $pS\frac{\partial V}{\partial S} + qV$ on I_{S_i} and $\nu = \nu_j$ by solving the following two-point boundary value problem

$$\left(p_j S \frac{\partial W}{\partial S} + q_{i+\frac{1}{2},j} W \right)' = 0, \quad W(S_i, \nu_j) = V_{i,j}, \quad W(S_{i+1}, \nu_j) = V_{i+1,j}, \tag{3.2.4}$$

where $p_j = p(\nu_j)$, $q_{i+\frac{1}{2},j} = q(x_{i+\frac{1}{2}}, y_j)$, and $V_{i,j}$ and $V_{i+1,j}$ are approximations to the nodal values of V at (S_i, ν_j) and (S_{i+1}, ν_j), respectively. Using the same argument as that for (1.3.7), we have

$$\left(pS \frac{\partial V}{\partial S} + qV \right)_{(S_{i+\frac{1}{2}}, \nu_j)} \approx q_{i+\frac{1}{2},j} \frac{S_{i+1}^{\alpha_{i,j}} V_{i+1,j} - S_i^{\alpha_{i,j}} V_{i,j}}{S_{i+1}^{\alpha_{i,j}} - S_i^{\alpha_{i,j}}} =: \tau_{i,j}(V), \tag{3.2.5}$$

where $\alpha_{i,j} = q_{i+\frac{1}{2},j}/p_j$. Therefore, we define the following approximation:

$$J_{i,j}^1 \approx S_{i+\frac{1}{2}} \left(\tau_{i,j} + d_j \frac{V_{i,j+1} - V_{i,j}}{h_{\nu_j}} \right) h_{\nu_j}. \tag{3.2.6}$$

Similarly, we approximate $pS\frac{\partial V}{\partial S} + qV$ on $I_{S_{i-1}}$ using (3.2.5) with i replaced with $i - 1$, yielding the following approximation to $J_{i,j}^2$:

$$J_{i,j}^2 \approx S_{i-\frac{1}{2}} \left(\tau_{i-1,j} + d_j \frac{V_{i,j+1} - V_{i,j}}{h_{\nu_j}} \right) h_{\nu_j}. \tag{3.2.7}$$

Case 2. Approximation of J_{1j}^1 and J_{1j}^2.

Note that (3.2.6) also holds true for J_{1j}^1. However, as mentioned in Sect. 1.3.1, (3.2.4) becomes degenerate on I_{S_0}. Thus, we need to approximate $pS\frac{\partial V}{\partial S} + qV$ on I_{S_0} in a different way. More specifically, following **Case II** in Sect. 1.3.1, we consider the following problem.

$$\left(p_j S \frac{\partial W}{\partial S} + q_{\frac{1}{2},j} W\right)' = C, \quad \lim_{S \to +0} W(S, v_j) = V_{0,j}, \quad W(S_1, v_j) = V_{1,j}, \quad (3.2.8)$$

where C is an additive constant. Solving (3.2.8) exactly and following the same argument for (1.3.12), we have the following approximation:

$$\left(pS\frac{\partial V}{\partial S} + qV\right)_{(S_{\frac{1}{2}}, v_j)} \approx \frac{1}{2}\left[(p_j + q_{\frac{1}{2},j})V_{1,j} - (p_j - q_{\frac{1}{2},j})V_{0,j}\right] =: \tau_{0,j}(V).$$

$$(3.2.9)$$

Similarly to (3.2.7), we define the following approximation to $J_{0,j}^2$.

$$J_{1,j}^2 \approx S_{\frac{1}{2}}\left(\tau_{0,j} + d_j \frac{V_{1,j+1} - V_{1,j}}{h_{v_j}}\right) h_{v_j}. \qquad (3.2.10)$$

To approximation of J_{ij}^3 and J_{ij}^4, we first write their common integrand as

$$a_{21}\frac{\partial V}{\partial S} + a_{22}\frac{\partial V}{\partial v} + b_2 V = v\left(\bar{p}(S)v\frac{\partial V}{\partial v} + \bar{q}(S, v)V + \bar{d}(S, v)\frac{\partial V}{\partial S}\right),$$

where $\bar{p} = \frac{1}{2}\sigma_v$, $\bar{q} = \mu_v - \frac{1}{2}\rho\sigma_v v^{1/2} - \sigma_v^2$ and $\bar{d} = \rho\sigma_v S v^{1/2}$. Then, consider the following 2-point boundary value problem on (v_j, v_{j+1}):

$$\left(\bar{p}_i v \frac{\partial W}{\partial v} + \bar{q}_{i,j+\frac{1}{2}} W\right)' = 0, \quad W(S_i, v_j) = V_{i,j}, \quad W(S_i, v_{j+1}) = V_{i,j+1},$$

where $\bar{p}_i = \bar{p}(S_i)$ and $\bar{q}_{i,j+\frac{1}{2}} = \bar{q}(S_i, v_{j+\frac{1}{2}})$. Using an argument similar to that in **Case 1** above, we define the following approximation:

$$\left(\bar{p}v\frac{\partial V}{\partial v} + \bar{q}V\right)_{(S_i, v_{j+\frac{1}{2}})} \approx \bar{q}_{i,j+\frac{1}{2}} \frac{v_{j+1}^{\bar{\alpha}_{i,j}} V_{i,j+1} - v_j^{\bar{\alpha}_{i,j}} V_{i,j}}{v_{j+1}^{\bar{\alpha}_{i,j}} - v_j^{\bar{\alpha}_{i,j}}} =: \bar{\tau}_{i,j}(V) \qquad (3.2.11)$$

for $i = 1, 2, \ldots, N_S - 1$ and $j = 0, 1, \ldots, N_v - 1$, where $\bar{\alpha}_{i,j} = \bar{q}_{i,j+\frac{1}{2}}/\bar{p}_i$. Thus, J_{ij}^1 and J_{ij}^2 are approximated respectively by

$$J_{i,j}^3 \approx v_{j+\frac{1}{2}}\left(\bar{\tau}_{i,j} + \bar{d}_{i,j}\frac{V_{i+1,j} - V_{i,j}}{h_{S_i}}\right)h_{S_i}, \; J_{i,j}^4 \approx v_{j-\frac{1}{2}}\left(\bar{\tau}_{i,j-1} + \bar{d}_{i,j}\frac{V_{i+1,j} - V_{i,j}}{h_{S_i}}\right)h_{S_i},$$

$$(3.2.12)$$

for $j = 1, 2, \ldots, N_v - 1$, where $\bar{d}_{i,j} = \bar{d}(S_i, v_j)$. Note **Case 2** above does not apply to the approximation of J_{ij}^3 and J_{ij}^4, as $v_0 = v_{\min} > 0$.

Replacing the integrals on the RHS of (3.2.3) with their respective approximations in (3.2.6)–(3.2.12), we have the following system approximating (3.2.2):

$$-\frac{\partial V_{i,j}}{\partial t}|\Omega_{ij}| + e_{i,j;i,j-1}V_{i,j-1} + e_{i,j;i-1,j}V_{i-1,j} + e_{i,j;i,j}V_{i,j}$$

$$+ e_{i,j;i+1,j}V_{i+1,j} + e_{i,j;i,j+1}V_{i,j+1} = 0, \quad (3.2.13)$$

for $i = 1, 2, \ldots, N_S - 1, \; j = 1, 2, \ldots, N_v - 1$, where

$$e_{1,j;1,j-1} = -h_{S_1}v_{j-\frac{1}{2}}\bar{q}_{1,j-\frac{1}{2}}\frac{v_{j-1}^{\bar{\alpha}_{1,j-1}}}{v_j^{\bar{\alpha}_{1,j-1}} - v_{j-1}^{\bar{\alpha}_{1,j-1}}}, \quad e_{1,j;i-1,j} = -\frac{1}{2}S_{\frac{1}{2}}h_{v_j}(p_j - q_{\frac{1}{2},j}), \quad (3.2.14)$$

$$e_{1,j;1,j} = S_{\frac{3}{2}}h_{v_j}\frac{q_{\frac{3}{2},j}S_1^{\alpha_{1,j}}}{S_2^{\alpha_{1,j}} - S_1^{\alpha_{1,j}}} + +h_{S_1}\left(\frac{\bar{q}_{1,j+\frac{1}{2}}v_{j+\frac{1}{2}}v_j^{\bar{\alpha}_{1,j}}}{v_{j+1}^{\bar{\alpha}_{1,j}} - v_j^{\bar{\alpha}_{1,j}}} + v_{j-\frac{1}{2}}\bar{q}_{1,j-\frac{1}{2}}\frac{v_j^{\bar{\alpha}_{1,j-1}}}{v_j^{\bar{\alpha}_{1,j-1}} - v_{j-1}^{\bar{\alpha}_{1,j-1}}}\right)$$

$$+ \frac{1}{2}S_{\frac{1}{2}}v_{y_j}(p_j + q_{\frac{1}{2},j}) + h_{v_j}\bar{d}_{1,j} + d_j h_{S_1} + c_{1,j}|\Omega_{ij}|, \quad (3.2.15)$$

$$e_{1,j;2,j} = -\bar{d}_{1,j}h_{v_j} - S_{\frac{3}{2}}h_{v_j}\frac{q_{\frac{3}{2},j}S_2^{\alpha_{1,j}}}{S_2^{\alpha_{1,j}} - S_1^{\alpha_{1,j}}}, \quad e_{1,j;1,j+1} = -d_j h_{S_1} - h_{S_1}\frac{\bar{b}_{1,j+\frac{1}{2}}v_{j+\frac{1}{2}}v_{j+1}^{\bar{\alpha}_{1,j}}}{S_{j+1}^{\bar{\alpha}_{1,j}} - v_j^{\bar{\alpha}_{1,j}}}$$

$$(3.2.16)$$

for $j = 1, 2, \ldots, N_y - 1$, and

$$e_{i,j;i,j-1} = -h_{S_i}v_{j-\frac{1}{2}}\bar{b}_{i,j-\frac{1}{2}}\frac{v_{j-1}^{\bar{\alpha}_{i,j-1}}}{v_j^{\bar{\alpha}_{i,j-1}} - v_{j-1}^{\bar{\alpha}_{i,j-1}}}, \quad e_{i,j;i+1,j} = -S_{i-\frac{1}{2}}h_{v_j}\frac{q_{i-\frac{1}{2},j}S_{i-1}^{\alpha_{i-1,j}}}{S_i^{\alpha_{i-1,j}} - S_{i-1}^{\alpha_{i-1,j}}},$$

$$(3.2.17)$$

$$e_{i,j;i,j} = h_{v_j}\left(S_{i+\frac{1}{2}}\frac{q_{i+\frac{1}{2},j}S_i^{\alpha_{i,j}}}{S_{i+1}^{\alpha_{i,j}} - S_i^{\alpha_{i,j}}} + S_{i-\frac{1}{2}}\frac{q_{i-\frac{1}{2},j}S_i^{\alpha_{i-1,j}}}{S_i^{\alpha_{i-1,j}} - S_{i-1}^{\alpha_{i-1,j}}}\right) + h_{v_j}\bar{d}_{i,j} + d_j h_{S_i}$$

$$+ h_{S_i}\left(\frac{\bar{b}_{i,j+\frac{1}{2}}v_{j+\frac{1}{2}}y_j^{\bar{\alpha}_{i,j}}}{v_{j+1}^{\bar{\alpha}_{i,j}} - v_j^{\bar{\alpha}_{i,j}}} + v_{j-\frac{1}{2}}\bar{q}_{i,j-\frac{1}{2}}\frac{v_j^{\bar{\alpha}_{i,j-1}}}{v_j^{\bar{\alpha}_{i,j-1}} - v_{j-1}^{\bar{\alpha}_{i,j-1}}}\right) + c_{i,j}|\Omega_{ij}|, \quad (3.2.18)$$

$$e_{i,j;i+1,j} = -\bar{d}_{i,j}h_{v_j} - S_{i+\frac{1}{2}}h_{v_j}\frac{q_{i+\frac{1}{2},j}S_{i+1}^{\alpha_{i,j}}}{S_{i+1}^{\alpha_{i,j}} - S_i^{\alpha_{i,j}}}, \quad e_{i,j;i,j+1} = -d_j h_{S_i} - h_{S_i}\frac{\bar{q}_{i,j+\frac{1}{2}}v_{j+\frac{1}{2}}v_{j+1}^{\bar{\alpha}_{i,j}}}{v_{j+1}^{\bar{\alpha}_{i,j}} - v_j^{\bar{\alpha}_{i,j}}},$$

$$(3.2.19)$$

for $i = 2, 3, \ldots, N_S - 1, j = 1, 2, \ldots, N_v - 1$.

Equation (3.2.13) is a linear system in $V_{i,j}, i = 0, 1, \ldots, N_S$ and $j = 0, 1, \ldots, N_v$. However, $V_{i,j}$ equal to the boundary values when $i = 0, i = N_S, j = 0$ or $j = N_v$. Thus, these known terms are cast to the RHS of (3.2.13). Rearrange the 2D array $\{V_{i,j}\}$ as a 1D one in the order $V_{1,1}, V_{2,1}, \ldots, V_{N_S-1,1}, \ldots, V_{1,N_v-1}, \ldots, V_{N_S-1,N_v-1}$. This mapping is achieved under the transformation $n = i + (j-1)(N_S - 1)$ for $i = 1, 2, \ldots, N_S - 1$ and $j = 1, 2, \ldots, N_v - 1$. Let $\bar{V}_n = V_{i,j}$ for a feasible (i, j). Then, (3.2.13) is an $N \times N$ linear system in $\bar{V} = (\bar{V}_1, \ldots, \bar{V}_N)^\top$, which can be written as

$$-\frac{\partial \bar{V}_n(t)}{\partial t}|\Omega_n| + \bar{E}_n \bar{V} = f_n \qquad (3.2.20)$$

for $n = 1, 2, \ldots, N$, where $\bar{E}_n = (\bar{e}_{n,1}, \ldots, \bar{e}_{n,n})^\top$ and f_n is the contribution from the boundary conditions g_1, g_2, G_1 and G_2 in (3.1.10)–(3.1.11). From our analysis we see that $\bar{E}_n$ has up to 5 non-zero entries given by $\bar{e}_{n,n-(N_S-1)} = e_{i,j;i,j-1}, \bar{e}_{n,n-1} = e_{i,j;i-1,j}, \bar{e}_{n,n} = e_{i,j;i,j}, \bar{e}_{n,n+1} = e_{i,j;i+1,j}$, and $\bar{e}_{n,n+(N_S-1)} = e_{i,j;i,j+1}$ with $e_{i,j;l,m}$ defined in (3.2.14)–(3.2.19). Other entries of $\bar{E}_n$ are zeros. The RHS term f_n is nonzero only when its corresponding mesh node index (i, j) satisfies $i = 1, i = N_S - 1, j = 1$ or $j = N_v - 1$.

Equation (3.2.20) is semi-discrete. To discretize it in t, we choose a mesh for $[0, T]$ with nodes $t_k (k = 0, 1, \ldots K)$ satisfying $T = t_0 > t_1 > \cdots > t_K = 0$. As in Sect. 1.3.2, the implicit time-stepping scheme with a splitting parameter $\theta \in [\frac{1}{2}, 1]$ is used to discretize the time derivative in (3.2.20), yielding

$$\frac{\bar{V}_n^{k+1} - \bar{V}_n^k}{-\Delta t_k}|\Omega_n| + \theta \bar{E}_n^{k+1} \bar{V}^{k+1} + (1 - \theta)\bar{E}_n^k \bar{V}^k = \theta f_n^{k+1} + (1 - \theta)f_n^k,$$

for $k = 0, 1, 2, \ldots, K - 1$ and $n = 1, 2, \ldots, N$ with $\bar{V}^0$ the payoff condition g_3 in (3.1.10), where $\Delta t_k = t_{k+1} - t_k < 0, \bar{E}_n^l = \bar{E}_n(t_l)$ and $f_n^l = f_n(t_l)$ for $l = k$ and $k + 1$. The above system can be written as the following matrix equation:

$$\begin{cases} (\theta \bar{E}^{k+1} + G^k)\bar{V}^{k+1} = [G^k - (1 - \theta)\bar{E}^k]\bar{V}^k + \theta f_n^{k+1} + (1 - \theta)f_n^k, \\ \bar{V}_n^0 = g_3(S_i, v_j), \ n = i + (j-1)(N_S - 1), \ i = 1, \ldots, N_S, \ j = 1, \ldots, N_v, \end{cases} \qquad (3.2.21)$$

for $k = 0, 1, \cdots, K - 1$, where $G^k = \frac{1}{-\Delta t_k}\text{diag}(|\Omega_1|, \ldots, |\Omega_N|)$. The cases that $\theta = \frac{1}{2}$ and $\theta = 1$ correspond to Crank–Nicolson and Backward Euler (or fully implicit) time-stepping schemes respectively. We now show that the system matrix of (3.2.21) is an M-matrix in the following theorem.

Theorem 3.2.1 Let $k \in \{0, 1, \ldots, K - 1\}$. If $|\Delta t_k|$ is sufficiently small, then the system matrix $\theta \bar{E}^{k+1} + G^k$ of (3.2.21) is an M-matrix.

Proof We have shown $\bar{E}^{k+1}$ is penta-diagonal with non-zero entries $\bar{e}_{n,n-(N_S-1)} = e_{i,j;i,j-1}$, $\bar{e}_{n,n-1} = e_{i,j;i-1,j}$, $\bar{e}_{n,n} = e_{i,j;i,j}$, $\bar{e}_{n,n+1} = e_{i,j;i+1,j}$, and $\bar{e}_{n,n+(N_S-1)} = e_{i,j;i,j+1}$ evaluated at t_k in the nth row, with $e_{i,j;l,m}$ defined in (3.2.14)–(3.2.19). We now show that $\bar{E}^{k+1}$ is an M-matrix when $c_{i,j}^{k+1} := c_{i,j}(t_{k+1}) \geq 0$.

From the proof of Theorems 1.3.1 and (1.3.8), we have that $\frac{\alpha}{S_{i+1}^\alpha - S_i^\alpha} > 0$ and $\frac{\alpha}{v_{j+1}^\alpha - v_j^\alpha} > 0$ for any $\alpha > 0$ and feasible (i, j). This, from their definitions we see that $e_{i,j;l,m}^{k+1} < 0$ if $i \neq l$ or $j \neq m$, and $e_{i,j;i,j}^{k+1} > 0$ when $c_{i,j}^{k+1} \geq 0$. Furthermore, from (3.2.14)–(3.2.19) we also have

$$e_{i,j;i,j}^{k+1} \geq |e_{i,j-1;i,j}^{k+1}| + |e_{i-1,j;i,j}^{k+1}| + |e_{i+1,j;i,j}^{k+1}| + |e_{i,j+1;i,j}^{k+1}| + c_{i,j}^{k+1}|\Omega_{ij}|,$$

since d_j^{k+1} and $\bar{d}_{i,j}^{k+1}$ are non-negative. Using the mapping $n = i + (j-1)(N_S - 1)$, we see that the above inequality is $\bar{e}_{n,n}^{k+1} \geq \sum_{1 \leq l \leq N, l \neq n} |\bar{e}_{j,n}^{n+1}|$, Therefore, $\bar{E}^{k+1}$ is diagonally dominant with respect to its columns when $c_{i,j}^{k+1} \geq 0$. Also, when $i = 1$, $i = N_S - 1$, $j = 1$ or $j = N_v - 1$, the corresponding row $n = i + (j-1)(N_S - 1)$ has only 4 or less non-zero off-diagonal entries (and $f_n \neq 0$). In this case, $\bar{e}_{n,n}^{k+1} > \sum_{1 \leq l \leq N, l \neq n} |\bar{e}_{j,n}^{n+1}|$. Obviously, $\bar{E}^{k+1}$ is irreducible, as otherwise, (3.2.21) can be decomposed into two disjoint sub-problems which can be solved sequentially. Thus, $\bar{E}^{k+1}$ is an irreducibly diagonally dominant matrix. Using the result from [10, p.85], we conclude that $\bar{E}^{k+1}$ is an M-matrix, when $c_{i,j}^{k+1} \geq 0$.

The matrix G^k in (3.2.21) is a diagonal matrix with positive diagonal entries. When $|\Delta t_k|$ is sufficiently small, we expect $c_{i,j}^{k+1} + \frac{|\Omega_{ij}|}{-\Delta t_k} > 0$. Therefore $\theta \bar{E}^{k+1} + G^k$ is an M-matrix when $|\Delta t_k|$ is sufficiently small, proving this theorem. $\square$

3.3 Convergence of the FVM

We present a convergence analysis for the above FVM, based on the discussion in [8]. For simplicity, we assume $\partial u / \partial t = 0$ in (3.1.16).

3.3.1 Reformulation of the FVM

For the dual meshes constructed in Sect. 1.2, we let $\psi_{i,j}$ denote the characteristic function given by $\psi_{i,j}(S, v) = 1$ when $(S, v) \in \Omega_{ij}$ ans 0 otherwise. We choose a test space $T_h := \text{span} \{\psi_{i,j}, i = 1, 2, \ldots, N_1 - 1, j = 1, 2, \ldots, N_2 - 1\}$.

To define the trial space, let

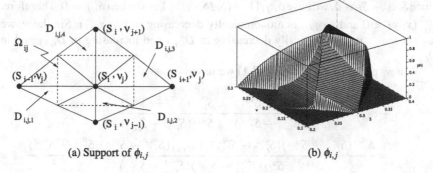

(a) Support of $\phi_{i,j}$ (b) $\phi_{i,j}$

Fig. 3.2 Hat function $\phi_{i,j}(S, v)$ and its support

$$
\phi_{i,j} = \begin{cases}
\left(\frac{S_i}{S}\right)^{\alpha_{i-1,j}} \left[1 - \left(\frac{S_i}{S_{i-1}}\right)^{\alpha_{i-1,j}}\right]^{-1} \left[1 - \left(\frac{S}{S_{i-1}}\right)^{\alpha_{i-1,j}}\right], & (S, v) \in D_{i,j,1}, \\
\left(\frac{S_i}{S}\right)^{\alpha_{i,j}} \left[1 - \left(\frac{S_i}{S_{i+1}}\right)^{\alpha_{i,j}}\right]^{-1} \left[1 - \left(\frac{S}{S_{i+1}}\right)^{\alpha_{i,j}}\right], & (S, v) \in D_{i,j,3}, \\
\left(\frac{v_j}{v}\right)^{\bar{\alpha}_{i,j-1}} \left[1 - \left(\frac{v_j}{v_{j-1}}\right)^{\bar{\alpha}_{i,j-1}}\right]^{-1} \left[1 - \left(\frac{v}{v_{j-1}}\right)^{\bar{\alpha}_{i,j-1}}\right], & (S, v) \in D_{i,j,2}, \\
\left(\frac{v_j}{v}\right)^{\bar{\alpha}_{i,j}} \left[1 - \left(\frac{v_j}{v_{j+1}}\right)^{\bar{\alpha}_{i,j}}\right]^{-1} \left[1 - \left(\frac{v}{v_{j+1}}\right)^{\bar{\alpha}_{i,j}}\right], & (S, v) \in D_{i,j,4}, \\
0, & \text{otherwise,}
\end{cases}
\tag{3.3.1}
$$

for $i = 2, 3, \ldots, N_S - 1$ and $j = 1, 2, \ldots, N_v - 1$, where $\alpha_{i,j}$ and $\bar{\alpha}_{i,j}$ are as defined in Sect. 3.2 and $D_{i,j,l}$, $l = 1, 2, 3, 4$, denote the quadrilateral domains depicted in Fig. 3.2a. For example $D_{i,j,1}$ is the diamond region with vertices (S_{i-1}, v_j), $(S_{i-\frac{1}{2}}, v_{j-\frac{1}{2}})$, (S_i, v_j) and $(S_{i-\frac{1}{2}}, v_{j+\frac{1}{2}})$. Each of the expressions in (3.3.1) is a solution to a local two-point boundary value problem such as (3.2.4). When $i = 1$, $\phi_{1,j} = \frac{S}{S_1}$ on $D_{1,j,1}$, which is a particular solution to (3.2.8). From (3.3.1) we see that the support of $\phi_{i,j}$ is $\cup_{l=1}^{4} D_{i,j,l}$ which is also a quadrilateral region as shown in Fig. 3.2a and an examples of $\phi_{i,j}$ is shown in Fig. 3.2b, from which it is seen that it is piecewise monotone and discontinuous across the inter-element boundaries. For these basis functions we have the following theorem:

Theorem 3.3.1 *The function in (3.3.1) has the following properties.*

1. $\phi_{i,j}$ is increasing on $D_{i,j,1}$ and $D_{i,j,2}$ and decreasing on $D_{i,j,3}$ and $D_{i,j,4}$.
2. $\phi_{i,j} + \phi_{i+1,j} = 1$ for $(S, v) \in D_{i,j,3}$ and $\phi_{i,j} + \phi_{i,j+1} = 1$ for $(S, v) \in D_{i,j,4}$.

Proof For Item 1, we only show $\phi_{i,j}$ is decreasing on $D_{i,j,3}$ in S. Differentiating $\phi_{i,j}$ with respect to S on $D_{i,j,3}$, we have

$$
\frac{\partial \phi_{i,j}}{\partial S}(S, v) = \frac{-\alpha_{i,j}}{1 - \left(\frac{S_i}{S_{i+1}}\right)^{\alpha_{i,j}}} \frac{S_i^{\alpha_{i,j}}}{S^{\alpha_{i,j}+1}}, \quad (S, v) \in D_{i,j,3}.
$$

Since $S_{i+1} - S_i > 0$, we have $\alpha_{i,j}/[1 - (S_i/S_{i+1})^{\alpha_{i,j}}] > 0$ when $\alpha_{i,j} \neq 0$. Therefore, $\frac{\partial \phi_{i,j}}{\partial x}(x, y) < 0$ and so $\phi_{i,j}$ is monotonically decreasing in (x_i, x_{i+1}). Similarly, we can show $\phi_{i,j}$ is monotonically decreasing in $D_{i,j,4}$ and increasing in $D_{i,j,1}$ and in $D_{i,j,2}$.

We now prove Item 2. From (3.3.1) we have

$$
\begin{aligned}
\phi_{i,j} + \phi_{i+1,j} &= \frac{S_i^{\alpha_{i,j}}(S_{i+1}^{\alpha_{i,j}} - S^{\alpha_{i,j}})}{S^{\alpha_{i,j}}(S_{i+1}^{\alpha_{i,j}} - S_i^{\alpha_{i,j}})} + \frac{S_{i+1}^{\alpha_{i,j}}(S_i^{\alpha_{i,j}} - S^{\alpha_{i,j}})}{S^{\alpha_{i,j}}(S_i^{\alpha_{i,j}} - S_{i+1}^{\alpha_{i,j}})} \\
&= \frac{S_i^{\alpha_{i,j}}(S_{i+1}^{\alpha_{i,j}} - S^{\alpha_{i,j}})(S_i^{\alpha_{i,j}} - S_{i+1}^{\alpha_{i,j}}) + S_{i+1}^{\alpha_{i,j}}(S_i^{\alpha_{i,j}} - S^{\alpha_{i,j}})(S_{i+1}^{\alpha_{i,j}} - S_i^{\alpha_{i,j}})}{S^{\alpha_{i,j}}(S_{i+1}^{\alpha_{i,j}} - S_i^{\alpha_{i,j}})(S_i^{\alpha_{i,j}} - S_{i+1}^{\alpha_{i,j}})} \\
&= \frac{(S_{i+1}^{\alpha_{i,j}} - S_i^{\alpha_{i,j}})(S_{i+1}^{\alpha_{i,j}} S_i^{\alpha_{i,j}} - S_{i+1}^{\alpha_{i,j}} S^{\alpha_{i,j}} - x_i^{\alpha_{i,j}} S_{i+1}^{\alpha_{i,j}} + S_i^{\alpha_{i,j}} S^{\alpha_{i,j}})}{S^{\alpha_{i,j}}(S_{i+1}^{\alpha_{i,j}} - S_i^{\alpha_{i,j}})(S_i^{\alpha_{i,j}} - S_{i+1}^{\alpha_{i,j}})} = 1
\end{aligned}
$$

for all $(S, v) \in D_{i,j,3}$. Similarly, $\phi_{i,j} + \phi_{i,j+1} = 1$, for all $(S, v) \in D_{i,j,4}$. $\square$

We choose the trial space $U_h = \operatorname{span}\{\phi_{i,j} : i = 1, \ldots, N_S - 1, j = 1, \ldots, N_v - 1\}$. Let P denote the mass lumping operator from $C^0(\bar{\Omega})$ to T_h defined by $P(v) = \sum_{i=0}^{N_S} \sum_{j=0}^{N_v} v(S_i, v_j)\psi_{i,j}$ and Q_1 and Q_2 denote other mass lumping operators from $C^1(\bar{\Omega})$ to T_h such that for any $v \in C^1(\bar{\Omega})$, $Q_1(\frac{\partial v}{\partial v}) = \sum_{i=0}^{N_S} \sum_{j=0}^{N_v - 1} \frac{v(S_i, v_{j+1}) - v(S_i, v_j)}{h_{v_j}} \psi_{i,j}$ and $Q_2(\frac{\partial v}{\partial S}) = \sum_{i=0}^{N_S - 1} \sum_{j=0}^{N_v} \frac{v(S_{i+1}, v_j) - v(S_i, v_j)}{h_{S_j}} \psi_{i,j}$ where $\psi_{i,j}$ is defined in before. Similarly to Problem 1.3.1, we define the following Petrov–Galerkin problem.

Problem 3.3.1 Find $u_h \in U_h$ such that for all $v_h \in T_h$

$$
A_h(u_h, v_h) = (P(f), v_h), \tag{3.3.2}
$$

where $A_h(\cdot, \cdot)$ denotes the bilinear form on $U_h \times T_h$ defined by

$$
\begin{aligned}
A_h(u_h, v_h) = &-\left(Q_1\left(d \frac{\partial u_h}{\partial v}\right), v_h\right) - \left(Q_2\left(\bar{d} \frac{\partial u_h}{\partial S}\right), v_h\right) + (P(\hat{c}u_h), v_h) \\
&- \sum_{i=1}^{N_S - 1} \sum_{j=1}^{N_v - 1} \left\{ S_{i+\frac{1}{2}} \tau_{i,j}(u_h) - S_{i-\frac{1}{2}} \tau_{i-1,j}(u_h) \right\} h_{v_j} v_h|_{\Omega_{ij}} \\
&- \sum_{i=1}^{N_S - 1} \sum_{j=1}^{N_v - 1} \left\{ v_{j+\frac{1}{2}} \bar{\tau}_{i,j}(u_h) - v_{j-\frac{1}{2}} \bar{\tau}_{i,j-1}(u_h) \right\} h_{S_i} v_h|_{\Omega_{ij}}, \tag{3.3.3}
\end{aligned}
$$

where $\tau_{i,j}(u)$ is defined in (3.2.5) and (3.2.9) and $\bar{\tau}_{i,j}$ is given in (3.2.11).

From the construction of the FVM, particularly the arguments leading to the approximation of $J_{i,j}^l$ for $l = 1, 2, 3, 4$, in Sect. 3.2 we see if $v_h = 1$ when $(S, v) \in \Omega_{ij}$ and 0 otherwise, then (3.3.2) becomes (3.2.13).

Using the lumping operator P, Problem 3.3.1 can be written as the following equivalent Galerkin finite element problem.

Problem 3.3.2 Find $u_h \in U_h$ such that for all $v_h \in U_h$,

$$B_h(u_h, v_h) := A_h(u_h, P(v_h)) = (P(f), v_h). \tag{3.3.4}$$

3.3.2 Stability and Convergence

For any $v_h \in U_h$, we define the following discrete norms and semi-norms.

$$\|v_h\|_{1,h_S}^2 := \sum_{i=1}^{N_S-1} \sum_{j=1}^{N_v-1} S_{i+1/2}^2 b_{i+1/2,j} h_{v_j} \frac{S_{i+1}^{\alpha_{i,j}} + S_i^{\alpha_{i,j}}}{S_{i+1}^{\alpha_{i,j}} - S_i^{\alpha_{i,j}}} (v_{i+1,j} - v_{i,j})^2,$$

$$\|v_h\|_{1,h_v}^2 := \sum_{i=1}^{N_S-1} \sum_{j=1}^{N_v-1} v_{j+1/2}^2 \bar{b}_{i,j+1/2} h_{S_i} \frac{v_{j+1}^{\bar{\alpha}_{i,j}} + v_j^{\bar{\alpha}_{i,j}}}{v_{j+1}^{\bar{\alpha}_{i,j}} - v_j^{\bar{\alpha}_{i,j}}} (v_{i,j+1} - v_{i,j})^2,$$

$$\|v_h\|_{0,h}^2 := \sum_{i=1}^{N_S-1} \sum_{j=1}^{N_v-1} v_{i,j}^2 |\Omega_{ij}|.$$

Using these norms and semi-norms, we define the following weighted discrete H^1-norm $\|v_h\|_h^2 = \|v_h\|_{1,h_S}^2 + \|v_h\|_{1,h_v}^2 + \|v_h\|_{0,h}^2$. The following theorem establishes the coercivity of $B_h(\cdot, \cdot)$.

Theorem 3.3.2 Let h be sufficiently small. For any $v_h \in U_h$, we have

$$B_h(v_h, v_h) \geq C \|v_h\|_h^2,$$

where C denotes the positive constant, independent of h and v_h.

Proof Let $v_h := \sum_{i=1}^{N_1-1} \sum_{j=1}^{N_2-1} v_{i,j} \phi_{i,j} \in U_h$. Replacing v_h in (3.3.3) with $P(v_h)$ gives

$$B_h(v_h, v_h) = -\sum_{i=1}^{N_S-1} \sum_{j=1}^{N_v-1} \left\{ S_{i+\frac{1}{2}} \tau_{i,j}(v_h) - S_{i-\frac{1}{2}} \tau_{i-1,j}(v_h) \right\} h_{v_j} v_{i,j}$$

$$- \sum_{i=1}^{N_S-1} \sum_{j=1}^{N_v-1} \left\{ v_{j+\frac{1}{2}} \bar{\tau}_{i,j}(v_h) - v_{j-\frac{1}{2}} \bar{\tau}_{i,j-1} \right\} h_{S_i} v_{i,j} + \sum_{i=1}^{N_S-1} \sum_{j=1}^{N_v-1} \frac{d_{i,j}}{h_{v_j}} (v_{i,j}^2 - v_{i,j+1}v_{i,j}) |\Omega_{ij}|$$

$$+ \sum_{i=1}^{N_S-1} \sum_{j=1}^{N_v-1} \frac{\bar{d}_{i,j}}{h_{S_i}} (v_{i,j}^2 - v_{i+1,j}v_{i,j}) |\Omega_{ij}| + \sum_{i=1}^{N_S-1} \sum_{j=1}^{N_v-1} \hat{c}_{i,j} v_{i,j}^2 |\Omega_{ij}|.$$

Rearranging the first and second sums with $v_{0,j} = 0$, we have

$$B(v_h, v_h) = \sum_{j=1}^{N_v-1} \frac{S_{1/2}(p_j + q_{\frac{1}{2},j})h_{v_j}}{2} v_{1,j}^2$$

$$-\sum_{i=1}^{N_S-1}\sum_{j=1}^{N_v-1} S_{i+\frac{1}{2}} \left(q_{i+\frac{1}{2},j} \frac{S_{i+1}^{\alpha_{i,j}} v_{i+1,j} - S_i^{\alpha_{i,j}} v_{i,j}}{S_{i+1}^{\alpha_{i,j}} - S_i^{\alpha_{i,j}}} \right) \cdot h_{v_j} v_{i,j}$$

$$+\sum_{i=2}^{N_S-1}\sum_{j=1}^{N_v-1} S_{i-\frac{1}{2}} \left(q_{i-\frac{1}{2},j} \frac{S_i^{\alpha_{i-1,j}} v_{i,j} - S_{i-1}^{\alpha_{i-1,j}} v_{i-1,j}}{S_i^{\alpha_{i-1,j}} - S_{i-1}^{\alpha_{i-1,j}}} \right) \cdot h_{v_j} v_{i,j}$$

$$-\sum_{i=1}^{N_S-1}\sum_{j=1}^{N_v-1} v_{j+\frac{1}{2}} \left(\bar{q}_{i,j+1/2} \frac{v_{j+1}^{\bar{\alpha}_{i,j}} v_{i,j+1} - v_j^{\bar{\alpha}_{i,j}} v_{i,j}}{v_{j+1}^{\bar{\alpha}_{i,j}} - v_j^{\bar{\alpha}_{i,j}}} \right) \cdot h_{S_i} v_{i,j}$$

$$+\sum_{i=1}^{N_S-1}\sum_{j=1}^{N_v-1} v_{j-1/2} \left(\bar{q}_{i,j-\frac{1}{2}} \frac{v_j^{\bar{\alpha}_{i,j-1}} v_{i,j} - v_{j-1}^{\bar{\alpha}_{i,j-1}} v_{i,j-1}}{v_j^{\bar{\alpha}_{i,j-1}} - v_{j-1}^{\bar{\alpha}_{i,j-1}}} \right) \cdot h_{S_i} v_{i,j}$$

$$+\sum_{i=1}^{N_S-1}\sum_{j=1}^{N_v-1} \frac{d_{i,j}}{h_{v_j}} (v_{i,j}^2 - v_{i,j+1}v_{i,j})|\Omega_{ij}| + \sum_{i=1}^{N_S-1}\sum_{j=1}^{N_v-1} \frac{\bar{d}_{i,j}}{h_{S_i}} (v_{i,j}^2 - v_{i+1,j}v_{i,j})|\Omega_{ij}|$$

$$+\sum_{i=1}^{N_S-1}\sum_{j=1}^{N_v-1} \hat{c}_{i,j} v_{i,j}^2 |\Omega_{ij}|$$

$$= \sum_{j=1}^{N_v-1} \frac{S_{\frac{1}{2}}(p_j + q_{\frac{1}{2},j})h_{v_j}}{2} v_{1,j}^2 + \sum_{i=1}^{N_S-1} v_{\frac{1}{2}} \bar{q}_{i,\frac{1}{2}} h_{S_i} \frac{v_1^{\bar{\alpha}_{i,0}}}{v_1^{\bar{\alpha}_{i,0}} - v_0^{\bar{\alpha}_{i,0}}} v_{i,1}^2$$

$$+\underbrace{\sum_{i=1}^{N_S-1}\sum_{j=1}^{N_v-1} S_{i+\frac{1}{2}} q_{i+\frac{1}{2},j} h_{v_j} \frac{S_{i+1}^{\alpha_{i,j}} v_{i+1,j} - S_i^{\alpha_{i,j}} v_{i,j}}{S_{i+1}^{\alpha_{i,j}} - S_i^{\alpha_{i,j}}} (v_{i+1,j} - v_{i,j})}_{I_1}$$

$$+\underbrace{\sum_{i=1}^{N_S-1}\sum_{j=1}^{N_v-1} v_{j+1/2} \bar{q}_{i,j+\frac{1}{2}} h_{S_i} \frac{v_{j+1}^{\bar{\alpha}_{i,j}} v_{i,j+1} - v_j^{\bar{\alpha}_{i,j}} v_{i,j}}{v_{j+1}^{\bar{\alpha}_{i,j}} - v_j^{\bar{\alpha}_{i,j}}} \cdot (v_{i,j+1} - v_{i,j})}_{I_2}$$

$$+\underbrace{\sum_{i=1}^{N_S-1}\sum_{j=1}^{N_v-1} \frac{d_{i,j}}{h_{v_j}} (v_{i,j}^2 - v_{i,j+1}v_{i,j})|\Omega_{ij}| + \sum_{i=1}^{N_S-1}\sum_{j=1}^{N_v-1} \frac{\bar{d}_{i,j}}{h_{S_i}} (v_{i,j}^2 - v_{i+1,j}v_{i,j})|\Omega_{ij}|}_{I_3}$$

$$+\sum_{i=1}^{N_S-1}\sum_{j=1}^{N_v-1} \hat{c}_{i,j} v_{i,j}^2 |\Omega_{ij}|. \tag{3.3.5}$$

Using the argument for simplifying the term I in (1.3.31), we see the terms I_1 and I_2 can be rewritten as follows respectively.

$$I_1 = \frac{1}{2}\|v_h\|_{1,h_S}^2 - \frac{1}{2}\sum_{i=1}^{N_S-1}\sum_{j=1}^{N_v-1}\left(S_{i-\frac{1}{2}}\frac{q_{i+\frac{1}{2},j}-q_{i-\frac{1}{2},j}}{h_{S_i}} + q_{i+\frac{1}{2},j}\right)v_{i,j}^2|\Omega_{ij}| - \frac{1}{2}\sum_{j=1}^{N_2-1}S_{\frac{1}{2}}q_{\frac{1}{2},j}h_{v_j}v_{1,j}^2,$$

$$(3.3.6)$$

$$I_2 = \frac{1}{2}\|v_h\|_{1,h_v}^2 - \frac{1}{2}\sum_{i=1}^{N_S-1}\sum_{j=1}^{N_v-1}\left(v_{j-\frac{1}{2}}\frac{\bar{q}_{i,j+\frac{1}{2}}-\bar{q}_{i,j-\frac{1}{2}}}{h_{v_j}} + \bar{q}_{i,j+\frac{1}{2}}\right)v_{i,j}^2|\Omega_{ij}| - \frac{1}{2}\sum_{i=1}^{N_S-1}v_{\frac{1}{2}}\bar{q}_{i,\frac{1}{2}}h_{S_i}v_{i,1}^2,$$

$$(3.3.7)$$

since $v_{N_S,j} = 0$ and $v_{i,N_v} = 0$. For the term I_3, we have

$$I_3 = \sum_{i=1}^{N_S-1}\sum_{j=1}^{N_v-1}\frac{d_{i,j}}{h_{v_j}}(v_{i,j}^2 - v_{i,j+1}v_{i,j})|\Omega_{ij}| + \sum_{i=1}^{N_1-1}\sum_{j=1}^{N_2-1}\frac{\bar{d}_{i,j}}{h_{x_i}}(v_{i,j}^2 - v_{i+1,j}v_{i,j})|\Omega_{ij}|$$

$$\geq -C\sum_{i=1}^{N_S-1}\sum_{j=1}^{N_v-1}v_{i,j}^2|\Omega_{ij}|,$$

$$(3.3.8)$$

where C denotes a generic positive constant, independent of h_S and h_v.

Using (3.3.6)–(3.3.8), we have from (3.3.5)

$$B(v_h,v_h) = \sum_{j=1}^{N_v-1}\frac{S_{\frac{1}{2}}(p_j+q_{\frac{1}{2},j})h_{v_j}}{2}v_{1,j}^2 + \sum_{i=1}^{N_S-1}v_{\frac{1}{2}}\bar{q}_{i,\frac{1}{2}}h_{S_i}\frac{v_1^{\bar{\alpha}_{i,0}}}{v_1^{\bar{\alpha}_{i,0}}-v_0^{\bar{\alpha}_{i,0}}}v_{i,1}^2 + \frac{1}{2}\|v_h\|_{1,h_S}^2 + \frac{1}{2}\|v_h\|_{1,h_v}^2$$

$$+ \sum_{i=1}^{N_S-1}\sum_{j=1}^{N_v-1}\left(\hat{c}_{i,j} - \frac{S_{i-\frac{1}{2}}}{2}\frac{q_{i+\frac{1}{2},j}-q_{i-\frac{1}{2},j}}{h_{S_i}} - \frac{q_{i+\frac{1}{2},j}}{2} - \frac{v_{j-\frac{1}{2}}}{2}\frac{\bar{q}_{i,j+\frac{1}{2}}-\bar{q}_{i,j-\frac{1}{2}}}{h_{v_j}} - \frac{\bar{q}_{i,j+\frac{1}{2}}}{2}\right)v_{i,j}^2|\Omega_{ij}|$$

$$- \frac{1}{2}\sum_{j=1}^{N_v-1}S_{\frac{1}{2}}q_{\frac{1}{2},j}h_{v_j}v_{1,j}^2 - \frac{1}{2}\sum_{i=1}^{N_S-1}v_{1/2}\bar{q}_{i,\frac{1}{2}}h_{S_i}v_{i,1}^2 + I_3$$

$$\geq \sum_{j=1}^{N_v-1}\frac{S_{\frac{1}{2}}p_j}{2}h_{v_j}v_{1,j}^2 + \frac{1}{2}\|v_h\|_{1,h_S}^2 + \frac{1}{2}\|v_h\|_{1,h_v}^2 + \sum_{i=1}^{N_S-1}\sum_{j=1}^{N_v-1}\left(\beta + c_{i,j} - \frac{S_{i-\frac{1}{2}}}{2}\frac{q_{i+\frac{1}{2},j}-q_{i-\frac{1}{2},j}}{h_{S_i}}\right.$$

$$\left. - \frac{q_{i+\frac{1}{2},j}}{2} - \frac{v_{j-\frac{1}{2}}}{2}\frac{\bar{q}_{i,j+\frac{1}{2}}-\bar{q}_{i,j-\frac{1}{2}}}{h_{v_j}} - \frac{\bar{q}_{i,j+\frac{1}{2}}}{2} - K\right)v_{i,j}^2|\Omega_{ij}|$$

$$\geq C\left(\|v_h\|_{1,h_S}^2 + \|v_h\|_{1,h_v}^2 + \sum_{i=1}^{N_S-1}\sum_{j=1}^{N_v-1}v_{i,j}^2|\Omega_{ij}|\right) \geq C\|v_h\|_h^2,$$

when h_S and h_v are sufficiently small and β is sufficiently large, completing the proof. $\square$

Theorem 3.3.2 implies that the FVM in Sect. 3.2 is numerically stable. Let $\tau(u) := pS\frac{\partial u}{\partial S} + qu$ and $\bar{\tau}(u) := \bar{p}v\frac{\partial u}{\partial v} + \bar{q}u$. Analogous to (1.3.14) or the expression in Lemma 1.3.1, we define the constant approximations to τ_h and $\bar{\tau}_h$ respectively by

$$\tau_h(u) = \begin{cases} \tau_{i,j}(u) & \text{if } (S, v) \in D_{i,j,3}, \ i = 0, \ldots, N_S - 1, j = 0, \ldots, N_v - 1, \\ 0 & \text{otherwise,} \end{cases}$$

$$\bar{\tau}_h(u) = \begin{cases} \bar{\tau}_{i,j}(u) & \text{if } (S, v) \in D_{i,j,4}, \ i = 0, \ldots, N_S - 1, j = 0, \ldots, N_v - 1, \\ 0 & \text{otherwise,} \end{cases}$$

where $\tau_{i,j}$ and $\bar{\tau}_{i,j}$ are given in (3.2.5), (3.2.9) and (3.2.11), and $D_{i,j,k}$ are defined in Fig. 3.2a with $D_{i,j,3} = D_{i+1,j,1}$ and $D_{i,j,4} = D_{i,j+1,2}$.

For these flux approximations, we have the following lemma.

Lemma 3.3.1 *Let w be a sufficiently smooth function, and w_I be the U_h-interpolant of w. Then, there exists a constant $C > 0$, independent of h, such that*

$$\|\tau(w) - \tau_h(w_I)\|_{\infty, D_{i,j,3}} \leq C(\|\tau(w)\|_{1,\infty,D_{i,j,3}} + \|q\|_{1,\infty,D_{i,j,3}}\|w\|_{\infty,D_{i,j,3}})h_{S_i},$$

$$\|\bar{\tau}(w) - \bar{\tau}_h(w_I)\|_{\infty, D_{i,j,4}} \leq C(\|\bar{\tau}(w)\|_{1,\infty,RD_{i,j,4}} + \|\bar{b}\|_{1,\infty,D_{i,j,4}}\|w\|_{\infty,D_{i,j,4}})h_{y_j},$$

for $i = 0, 1, \ldots, N_1 - 1$ and $j = 0, 1, \ldots, N_2 - 1$.

The proof of this lemma is essentially the same as that of Lemma 1.3.2 and thus we omit it. Using this lemma we have our following theorem.

Theorem 3.3.3 *Let u and u_h the solutions to Problems 3.1.1 and 3.3.2 respectively. Then, there exists a constant C, independent of h, u_h and u, such that*

$$\|u_I - u_h\|_h \leq Ch(\|\tau(u)\|_{1,\infty} + \|\bar{\tau}(u)\|_{1,\infty} + \|u\|_{1,\infty} + \|q\|_{1,\infty} + \|\bar{q}\|_{1,\infty}$$
$$+ \|d\|_{1,\infty} + \|\bar{d}\|_{1,\infty} + \|\hat{c}\|_{1,\infty} + \|u\|_{2,\infty} + \|f\|_1), \tag{3.3.9}$$

where u_I is the U_h-interpolant of u.

Proof We assume that C is a generic positive constant, independent of h, u_h and u For any $v_h \in S_h$, multiplying (3.1.12) (with $\frac{\partial u}{\partial t} = 0$) by $P(v_h)$ and using integration by parts and (3.3.4), we have

$$-\sum_{i=1}^{N_S-1}\sum_{j=1}^{N_v-1}\left\{\left[S\tau(u) + d\frac{\partial u}{\partial v}\right]_{(S_{i-\frac{1}{2}},v_j)}^{(S_{i+\frac{1}{2}},v_j)}h_{v_j} + \left[v(\bar{\tau}(u)) + \bar{d}\frac{\partial u}{\partial S}\right]_{(S_i,v_{j-\frac{1}{2}})}^{(S_i,v_{j+\frac{1}{2}})}h_{S_i}\right\}P(v_h)$$
$$+ (\hat{c}u, P(v_h)) = (f, P(v_h)) = (f - P(f), P(v_h)) + B_h(u_h, v_h).$$

Taking $B_h(u_I, v_h)$ away from both side of the above equality and using an argument similar to that for (3.3.5), we have from the above equality,

$$
|B_h(u_h - u_I, v_h)| = \left| (\hat{c}u - P(\hat{c}u_I), P(v_h)) - \sum_{i=1}^{N_S-1} \sum_{j=1}^{N_v-1} [S(\tau(u) - \tau_h(u_I))]_{(S_{i-\frac{1}{2}}, v_j)}^{(S_{i+\frac{1}{2}}, v_j)} h_{v_j} v_{i,j} \right.
$$

$$
- \sum_{i=1}^{N_S-1} \sum_{j=1}^{N_v-1} [v(\bar{\tau}(u) - \bar{\tau}_h(u_I))]_{(S_i, v_{j-\frac{1}{2}})}^{(S_i, v_{j+\frac{1}{2}})} h_{S_i} v_{i,j} - \left(d\frac{\partial u}{\partial v} - Q_1\left(d\frac{\partial u_I}{\partial v} \right), P(v_h) \right)
$$

$$
\left. - \left(\bar{d}\frac{\partial u}{\partial S} - Q_2\left(\bar{d}\frac{\partial u_I}{\partial S} \right), P(v_h) \right) - (f - P(f), P(v_h)) \right|
$$

$$
\leq \underbrace{\sum_{i=1}^{N_S-1} \sum_{j=1}^{N_v-1} \int_{\Omega_{ij}} |\hat{c}u - \hat{c}_{i,j} u_{i,j}| \, d\Omega \, v_{i,j}}_{R_1}
$$

$$
+ \underbrace{\sum_{i=1}^{N_S-1} \sum_{j=1}^{N_v-1} v_{i,j} \int_{\Omega_{ij}} \left[\left| d\frac{\partial u}{\partial v} - d_{i,j} \frac{u_{i,j+1} - u_{i,j}}{h_{v_j}} \right| + \left| \bar{d}\frac{\partial u}{\partial S} - \bar{d}_{i,j} \frac{u_{i+1,j} - u_{i,j}}{h_{S_i}} \right| \right] d\Omega}_{R_2}
$$

$$
+ \underbrace{\left| \sum_{i=0}^{N_S-1} \sum_{j=1}^{N_v-1} [S_{i+\frac{1}{2}} (\tau(u) - \tau_h(u_I))_{(S_{i+\frac{1}{2}}, v_j)} (v_{i+1,j} - v_{i,j})] h_{v_j} \right|}_{R_3}
$$

$$
+ \underbrace{\left| \sum_{i=1}^{N_S-1} \sum_{j=0}^{N_v-1} [v_{j+\frac{1}{2}} (\bar{\tau}(u) - \bar{\tau}_h(u_I))_{(S_i, v_{j+\frac{1}{2}})} (v_{i,j+1} - v_{i,j})] h_{S_i} \right|}_{R_4} + \underbrace{|(f - P(f), P(v_h))|}_{R_5}.
$$

$$
(3.3.10)
$$

Since the mass lumping operator P preserves constants, it is easy to show that

$$
R_1 + R_5 \leq Ch(\|\hat{c}\|_{1,\infty} + \|u\|_{1,\infty} + \|f\|_1)\|v_h\|_h.
$$

For the first part of R_2, we have

$$
\sum_{i=1}^{N_S-1} \sum_{j=1}^{N_v-1} v_{i,j} \int_{\Omega_{ij}} \left| d\frac{\partial u}{\partial v} - d_{i,j} \frac{u_{i,j+1} - u_{i,j}}{h_{v_j}} \right| d\Omega \leq \sum_{i=1}^{N_S-1} \sum_{j=1}^{N_v-1} v_{i,j} \int_{\Omega_{ij}} \left| (d - d_{i,j}) \frac{\partial u}{\partial v} \right| d\Omega
$$

$$
+ \sum_{i=1}^{N_S-1} \sum_{j=1}^{N_v-1} v_{i,j} \int_{\Omega_{ij}} \left| d_{i,j} \left(\frac{\partial u}{\partial S} - \frac{u_{i,j+1} - u_{i,j}}{h_{v_j}} \right) \right| d\Omega \leq Ch(\|d\|_{q,\infty} + \|u\|_{2,\infty})\|v_h\|_h
$$

Similarly, we have estimate for the 2nd part of R_2. Thus, we conclude

$$
R_2 \leq Ch(\|d\|_{1,\infty} + \|\bar{d}\|_{1,\infty} + \|u\|_{2,\infty})\|v_h\|_h.
$$

Using Lemma 3.3.1 and Cauchy–Schwarz inequality we estimate R_3 as follows.

$$R_3 \leq Ch(\|\tau(u)\|_{1,\infty} + \|q\|_{1,\infty}\|u\|_{\infty}) \sum_{i=0}^{N_S-1} \sum_{j=1}^{N_v-1} S_{i+\frac{1}{2}} h_{v_j} |v_{i+1,j} - v_{i,j}|$$

$$\leq Ch(\|\tau(u)\|_{1,\infty} + \|q\|_{1,\infty}\|u\|_{\infty}) \left(S_1 \sum_{j=1}^{N_v-1} h_{v_j} |v_{1,j}| \right.$$

$$+ \sum_{i=1}^{N_S-1} \sum_{j=1}^{N_v-1} S_{i+\frac{1}{2}}^{\frac{1}{2}} \left[\frac{S_{i+1}^{\alpha_{i,j}} - S_i^{\alpha_{i,j}}}{q_{i+\frac{1}{2},j}(S_{i+1}^{\alpha_{i,j}} + S_i^{\alpha_{i,j}})} \right]^{\frac{1}{2}} \left. \frac{S_{i+\frac{1}{2}}^{\frac{1}{2}} q_{i+\frac{1}{2},j}^{\frac{1}{2}} h_{v_j} (S_{i+1}^{\alpha_{i,j}} + S_i^{\alpha_{i,j}})^{\frac{1}{2}}}{(S_{i+1}^{\alpha_{i,j}} - S_i^{\alpha_{i,j}})^{\frac{1}{2}}} |v_{i+1,j} - v_{i,j}| \right]$$

$$\leq Ch(\|\tau(u)\|_{1,\infty} + \|q\|_{1,\infty}\|u\|_{\infty}) \left[S_1 \sum_{j=1}^{N_v-1} h_{v_j} |v_{1,j}| + \left(\sum_{i=1}^{N_S-1} \sum_{j=1}^{N_v-1} \frac{S_{i+\frac{1}{2}}(S_{i+1}^{\alpha_{i,j}} - S_i^{\alpha_{i,j}})}{q_{i+\frac{1}{2},j}(S_{i+1}^{\alpha_{i,j}} + S_i^{\alpha_{i,j}})} \right)^{\frac{1}{2}} \right.$$

$$\cdot \left(\sum_{i=1}^{N_S-1} \sum_{j=1}^{N_v-1} q_{i+\frac{1}{2},j} S_{i+\frac{1}{2}} h_{v_j} \frac{S_{i+1}^{\alpha_{i,j}} + S_i^{\alpha_{i,j}}}{S_{i+1}^{\alpha_{i,j}} - S_i^{\alpha_{i,j}}} (v_{i+1,j} - v_{i,j})^2 \right)^{\frac{1}{2}} \right]$$

$$= Ch(\|\tau(u)\|_{1,\infty} + \|q\|_{1,\infty}\|u\|_{\infty}) \left[S_1 \sum_{j=1}^{N_v-1} h_{v_j} |v_{1,j}| + \left(\sum_{i=1}^{N_S-1} \sum_{j=1}^{N_v-1} \frac{S_{i+\frac{1}{2}}}{q_{i+\frac{1}{2},j}} \frac{S_{i+1}^{\alpha_{i,j}} - S_i^{\alpha_{i,j}}}{(S_{i+1}^{\alpha_{i,j}} + S_i^{\alpha_{i,j}})} \right)^{\frac{1}{2}} \|v_h\|_{1,h_S} \right].$$

$$(3.3.11)$$

By a Taylor expansion we have

$$\frac{S_{i+1}^{\alpha_{i,j}} - S_i^{\alpha_{i,j}}}{S_{i+1}^{\alpha_{i,j}} + S_i^{\alpha_{i,j}}} = \frac{\left(1 + \frac{h_{S_i}}{2x_{i+\frac{1}{2}}}\right)^{\alpha_{i,j}} - \left(1 - \frac{h_{S_i}}{2x_{i+\frac{1}{2}}}\right)^{\alpha_{i,j}}}{\left(1 + \frac{h_{S_i}}{2S_{i+\frac{1}{2}}}\right)^{\alpha_{i,j}} + \left(1 - \frac{h_{S_i}}{2x \S_{i+\frac{1}{2}}}\right)^{\alpha_{i,j}}}$$

$$= \frac{\left(1 + \alpha_{i,j}\mathcal{O}\left(\frac{h_{S_i}}{2S_{i+\frac{1}{2}}}\right)\right) - \left(1 - \alpha_{i,j}\mathcal{O}\left(\frac{h_{S_i}}{2S_{i+\frac{1}{2}}}\right)\right)}{\left(1 + \alpha_{i,j}\mathcal{O}\left(\frac{h_{x_i}}{2x_{i+1/2}}\right)\right) + \left(1 - \alpha_{i,j}\mathcal{O}\left(\frac{h_{x_i}}{2x_{i+1/2}}\right)\right)} \leq C\alpha_{i,j}\frac{h_{S_i}}{S_{i+\frac{1}{2}}}.$$

From this estimate, we see that the sum in (3.3.11) can be estimated as

$$\sum_{i=1}^{N_S-1} \sum_{j=1}^{N_v-1} \frac{S_{i+\frac{1}{2}}}{q_{i+\frac{1}{2},j}} \frac{S_{i+1}^{\alpha_{i,j}} - S_{i,j}^{\alpha}}{S_{i+1}^{\alpha_{i,j}} + S_i^{\alpha_{i,j}}} \leq C \sum_{i=1}^{N_S-1} \sum_{j=1}^{N_v-1} h_{S_i}\frac{\alpha_{i,j}}{q_{i+\frac{1}{2},j}} \leq C,$$

since $\alpha_{i,j}/q_{i+\frac{1}{2},j} = p_j$. Therefore,

$$R_3 \leq Ch(\|\tau(u)\|_{1,\infty} + \|q\|_{1,\infty}\|u\|_{\infty}) \left(\sum_{i=1}^{N_S-1} \sum_{j=1}^{N_v-1} |v_{1,j}| h_S h_v + \|v_h\|_{1,h_S} \right)$$

$$\leq Ch(\|\tau(u)\|_{1,\infty} + \|q\|_{1,\infty})\|v_h\|_h.$$

Since R_3 and R_4 are symmetric, by symmetry, we have $R_4 \leq Ch(\|\bar{\tau}(u)\|_{1,\infty} + \|\bar{q}\|_{1,\infty})\|v_h\|_h$.

Replacing R_i's in (3.3.10) with their respective bounds above, we obtain

$$|B(u_h - u_I, v_h)| \leq Ch(\|\tau(u)\|_{1,\infty} + \|\bar{\tau}(u)\|_{1,\infty} + \|u\|_{1,\infty} + \|q\|_{1,\infty} + \|\bar{q}\|_{1,\infty}$$
$$+ \|d\|_{1,\infty} + \|\bar{d}\|_{1,\infty} + \|\hat{c}\|_{1,\infty} + \|u\|_{2,\infty} + \|f\|_1)\|v_h\|_h.$$

Setting $v_h = u_h - u_I$ in the above estimate and using Theorem 3.3.2 we obtain (3.3.9). □

We remark that we can replace u_I in (3.3.9) with u, as the norm $\|\cdot\|_h$ only uses values at the mesh nodes and both u and u_I have the save nodal values. The following theorem shows that u_h depends continuously on the given data f.

Theorem 3.3.4 *Let u_h be the solution to Problem 3.3.4. There exists a constant $C > 0$, independent of h and u_h, such that $\|u_h\|_h \leq C\|f\|_{0,h}$.*

Proof Setting $v_h = u_h$ in (3.3.4) and using Cauchy–Schwarz inequality we have

$$B_h(u_h, u_h) = (P(f), P(u_h)) = \sum_{i=1}^{N_S-1} \sum_{j=1}^{N_v-1} u_{ij} f_{ij} |\Omega_{ij}| \leq \|u_h\|_{0,h} \|f\|_{0,h} \leq C\|u_h\|_h \|f\|_{0,h}.$$

Combining Theorem 3.3.2 and the above inequality yields $\|u_h\|_h \leq C\|f\|_{0,h}$. □

3.4 Power Penalty Method for Pricing American Options with Stochastic Volatility

In this section we shall extend the power penalty method in Chap. 2 to pricing American put options with stochastic volatility.

3.4.1 The Linear Complementarity Problem

Let V be the valuation of an American put option on one asset with stochastic volatility. We introduce $u = e^{\beta}(V_0 - V)$, where V_0 is a sufficiently smooth function satisfying the boundary conditions in (3.1.10) as aforementioned. As in Chap. 2, u is governed by the following Linear Complementarity Problem (LCP).

$$\mathscr{L}_{2D}u \leq f, \quad u - u^* \leq 0, \quad (\mathscr{L}_{2D}u - f)(u - u^*) = 0 \qquad (3.4.1)$$

in $\Omega \times [0, T)$ satisfying $u(S, v, t) = 0$ for $(S, v, t) \in \partial\Omega \times (0, T)$ and $u(S, v, T) = g(S, v)$, where $\mathscr{L}_{2D}$ is defined in (3.1.12) and u^* is a given lower bound on u. In what follows, we assume, as in (2.1.8), $u^* = e^{\beta t}(V_0(S, v) - V^*(S))$, where V^* is the Vanilla payoff function defined by $V^*(S) = \max\{0, K - S\}$.

Let $\mathcal{K} = \{v \in H^1_{0,w}(\Omega) : v \leq u^*\}$, where $H^1_{0,w}(\Omega)$ is defined Sect. 3.1.2. Similarly to Theorem 2.1.1, we have that following variational inequality which is the variational form of (3.4.1).

Problem 3.4.1 Find $u(t) \in \mathcal{K}$ such that, for all $v \in \mathcal{K}$,

$$\left(-\frac{\partial u}{\partial t}, v - u\right) + B(u, v - u; t) \geq (f, v - u) \qquad (3.4.2)$$

for $t \in (0, T)$ a.e., where $B(u, v; t)$ is a bilinear form defined in (3.1.15).

The unique solvability of Problem 3.4.1 is given in the following theorem.

Theorem 3.4.1 *Problem 3.4.2 has a unique solution.*

Proof In the proof of Theorem 3.1.1, we have shown $B(\phi, \phi; t) \geq C\|\phi\|^2_{1,w}$ and $B(\phi, \psi; t) \leq M\|\phi\|_{1,w}\|\psi\|_{1,w}$ for any $\phi, \psi \in H^1_{0,w}(\Omega)$. From [3, Lemma 1 & Theorem 1.33], we see that Problem 3.4.2 has a unique solution.   □

3.4.2 The Penalty Method and Convergence

Following the idea in Sect. 2.2.1, we approximate (3.4.1) by the following equation:

$$\mathcal{L}_{2D}u_\lambda + \lambda[u_\lambda - u^*]^{1/\kappa}_+ = f, \quad (S, v, t) \in \Omega \times [0, T] \qquad (3.4.3)$$

satisfying the following boundary and payoff conditions

$$u_\lambda(S, v, t) = 0 \text{ for } (S, v) \in \partial\Omega \quad \text{and} \quad u_\lambda(S, v, T) = u^*(S, , v, T), \qquad (3.4.4)$$

where $\lambda > 1$ and $\kappa > 0$ are parameters. The variational problem corresponding to (3.4.3)–(3.4.4) is given below.

Problem 3.4.2 Find $u_\lambda(t) \in H^1_{0,w}(\Omega)$ such that, for all $v \in H^1_{0,w}(\Omega)$,

$$\left(-\frac{\partial u_\lambda}{\partial t}, v\right) + B(u_\lambda, v; t) + \lambda\left([u_\lambda - u^*]^{1/\kappa}_+, v\right) = (f, v), \quad t \in (0, T) \text{ a.e..} \tag{3.4.5}$$

For this problem, we have the following theorem.

Theorem 3.4.2 *Problem 3.4.2 has a unique solution in $H^1_{0,w}(\Omega)$.*

The proof is identical to that of Theorem 3.4.2, and this it is omitted here. Before further discussion, we first introduce a space. Let H be a Hilbert space. we use $L^p(0, T; H(I))$ to denote the space defined by

$$L^p(0, T; H) = \{v : v((\cdot, \cdot, t)) \in H \text{ a.e. in } (0, T); \|v((\cdot, \cdot, t), t)\|_H \in L^p((0, T))\},$$

for a $p \in [1, \infty]$, equipped with the norm $\|v\|_{L^p(0,T;H)} = \left(\int_0^T \|v(\cdot, t)\|_H^p dt \right)^{1/p}$. Using this space, we have the following lemma.

Lemma 3.4.1 *Let u_λ be the solution to Problem 3.4.2. If $u_\lambda \in L^p(\Omega \times (0,T))$ with $p = 1 + 1/\kappa$, then there exists a constant $C > 0$, independent of u_λ and λ, such that*

$$\|[u_\lambda - u^*]_+\|_{L^p(\Omega \times (0,T))} \leq \frac{C}{\lambda^\kappa}, \tag{3.4.6}$$

$$\|[u_\lambda - u^*]_+\|_{L^\infty(0,T;L^2(\Omega))} + \|[u_\lambda - u^*]_+\|_{L^2(0,T;H^1_{0,w}(\Omega))} \leq \frac{C}{\lambda^{\kappa/2}}. \tag{3.4.7}$$

Proof Let C is a generic positive constant, independent of u_λ and λ. We introduce $\phi(S, v, t) := [u_\lambda - u^*]_+$. Note $u_\lambda \leq u^*$ is satisfied on $\partial\Omega$. This is because, when determine the boundary conditions for V, we either choose $V(S, v, t) = V^*(S, t)$ for $S = 0$ and $S = S_{max}$, or solve the 1D LCP (2.1.7) in Sect. 2.1.2 at $v = v_{min}$ and v_{max}. Thus $\phi(\cdot, \cdot, t) \in H^1_{0,w}(\Omega)$ a.e. in $(0, T)$. Replacing v with ϕ in (3.4.5) gives

$$\left(-\frac{\partial u_\lambda}{\partial t}, \phi \right) + B(u_\lambda, \phi; t) + \lambda(\phi^{1/k}, \phi) = (f, \phi), \quad \text{a.e. in } (0, T).$$

Subtracting $-(\frac{\partial u^*}{\partial t}, \phi) + B(u^*, \phi; t)$ from both sides of the above equation and then integrating the resulting equation from t to T, we have

$$\int_t^T \left(-\frac{\partial(u_\lambda - u^*)}{\partial \tau}, \phi \right) d\tau + \int_t^T B(u_\lambda - u^*, \phi; \tau)d\tau + \lambda \int_t^T (\phi^{1/k}, \phi)$$

$$= \int_t^T (f, \phi)d\tau + \int_t^T \left(\frac{\partial u^*}{\partial t}, \phi \right) d\tau - \int_t^T B(u^*, \phi; \tau)d\tau$$

$$\leq \left(\int_t^T \|f(\tau)\|_{L^q(\Omega)}^q d\tau \right)^{1/q} \left(\int_t^T \|\phi(\tau)\|_{L^p(\Omega)}^p d\tau \right)^{1/p}$$

$$+ \beta \int_t^T e^{\beta\tau}(V_0 - V^*, \phi(\tau))d\tau - \int_t^T B(u^*(\tau), \phi(\tau); \tau)d\tau, \tag{3.4.8}$$

where $q = 1 + \kappa$ so that $1/p + 1/q = 1$. In the above we used Hölder's inequality. Note $\frac{\partial(u_\lambda - u^*)}{\partial t} \cdot \phi = \frac{\partial\phi}{\partial t} \cdot \phi$. Using integration bu parts, we have

$$\int_t^T \left(-\frac{\partial(u_\lambda - u^*)}{\partial \tau}, \phi \right) d\tau = (\phi(t), \phi(t)) - \int_t^T \left(-\frac{\partial\phi}{\partial \tau}, \phi \right) d\tau,$$

since $\phi(\cdot, \cdot, T) = 0$. Thus,

$$\int_t^T \left(-\frac{\partial\phi}{\partial \tau}, \phi \right) d\tau = \frac{1}{2}(\phi(t), \phi(t)) = \frac{1}{2}\|\phi(t)\|_{L^2(\Omega)}^2. \tag{3.4.9}$$

From the above estimate, (3.4.8) and (3.1.17), we get

$$\frac{1}{2}(\phi, \phi) + C \int_t^T \|\phi\|_{1,w}^2 d\tau + \lambda \int_t^T \|\phi\|_{L^p(\Omega)}^p d\tau \leq C \left(\int_t^T \|\phi\|_{L^p(\Omega)}^p d\tau \right)^{1/p}$$
$$+ \beta \int_t^T e^{\beta\tau} (V_0 - V^*, \phi) \, d\tau - \int_t^T B(u^*, \phi) d\tau. \tag{3.4.10}$$

Let us consider the last two integrals in (3.4.10). Since $V_0(S, v)$ and $V^*(S)$ are bounded on $\bar{\Omega}$, using Hölder inequality we have

$$\int_t^T e^{\beta\tau} (V_0 - V^*, \phi) \, d\tau \leq C \int_t^T \int_\Omega \phi \, d\Omega \, d\tau \leq C \left(\int_t^T \|\phi\|_{L^p(\Omega)}^p d\tau \right)^{1/p}. \tag{3.4.11}$$

From (3.1.15) and u^* given in Sect. 3.4.1, we see that

$$-B(u^*, \phi; \tau) = -\left(A\nabla u^* + bu^*, \nabla\phi\right) + (\hat{c}u^*, \phi)$$
$$= e^{\beta t} \left[(A\nabla V_0 + bV_0, \nabla\phi) + (\hat{c}V_0, \phi) - \left(A\nabla V^* + bV^*, \nabla\phi\right) - (\hat{c}V^*, \phi) \right]. \tag{3.4.12}$$

Let $\Omega_1 = \{(S, v) \in \Omega : S < K\}$ and $\Omega_2 = \{(S, v) \in \Omega : S > K\}$. From the definition of V^* in Sect. 3.4.1, we have that $\nabla V^* = (K, 0)^\top$ when $(S, v) \in \Omega_1$ and $(0, 0)^\top$ when $(S, v) \in \Omega_2$. Therefore, integrating by parts gives

$$(A\nabla V^*, \nabla\phi) = \int_{\Omega_1} K(a_{11}, a_{21})^\top \cdot \nabla\phi \, dS \, dv = K \int_{\partial\Omega_1} \phi(a_{11}, a_{21})^\top \cdot \mathbf{n} \, ds$$
$$- K \int_{\Omega_1} \left(\frac{\partial a_{11}}{\partial S} + \frac{\partial a_{21}}{\partial v} \right) \phi \, d\Omega \leq C \int_\Omega \phi \, d\Omega \tag{3.4.13}$$

in $(0, T)$, because $a_{11}, a_{12} \geq 0$ and $\phi \geq 0$. Note $u^* \in H^1(\Omega)$ and $\phi \in H_{0,w}^1(\Omega)$. Using integration by parts again, we have

$$-(bu^*, \nabla\phi) + (\hat{c}u^*, \phi) = \left(\nabla \cdot (bu^*), \phi\right) + (\hat{c}u^*, \phi) \leq C \int_\Omega \phi \, d\Omega. \tag{3.4.14}$$

By a similar argument for (3.4.13) and (3.4.14), we are able to show

$$(A\nabla V_0 + bV_0, \nabla\phi) + (\hat{c}V_0, \phi) \leq C \int_\Omega \phi \, d\Omega. \tag{3.4.15}$$

Integrating (3.4.12) from t to T and using (3.4.13)–(3.4.15), we have

$$-\int_t^T B(u^*(\tau), \phi(\tau); \tau) d\tau \leq C \int_\Omega \phi(S, \tau) d\Omega \leq C \left(\int_t^T \|\phi(\tau)\|_{L^p(\Omega)}^p d\tau \right)^{1/p}.$$

Replacing the last two terms in (2.2.10) with (3.4.11) and the above upper bound respectively, we obtain

$$\frac{1}{2}(\phi(t), \phi(t)) + \int_t^T \|\phi\|_{1,w}^2 d\tau + \lambda \int_t^T \|\phi\|_{L^p(\Omega)}^p d\tau \leq C \left(\int_t^T \|\phi\|_{L^p(\Omega)}^p d\tau \right)^{1/p}$$
(3.4.16)

for all $t \in [0, T)$ a.e., which implies

$$\lambda \int_t^T \|\phi(\tau)\|_{L^p(\Omega)}^p d\tau \leq C \left(\int_t^T \|\phi(\tau)\|_{L^p(\Omega)}^p d\tau \right)^{1/p}.$$

Thus $\left(\int_t^T \|\phi(\tau)\|_{L^p(\Omega)}^p d\tau \right)^{1-1/p} \leq \frac{C}{\lambda}$, from which we obtain

$$\left(\int_t^T \|\phi(\tau)\|_{L^p(\Omega)}^p d\tau \right)^{1/p} \leq \frac{C}{\lambda^{1/(p-1)}} = \frac{C}{\lambda^\kappa},$$
(3.4.17)

since $p = 1 + 1/\kappa$. Thus, we have proved (3.4.6).

From (3.4.16) and (3.4.17), we have

$$\frac{1}{2}(\phi(t), \phi(t)) + \int_t^T \|\phi(\tau)\|_{1,w}^2 d\tau \leq C \left(\int_t^T \|\phi(\tau)\|_{L^p(\Omega)}^p d\tau \right)^{1/p} \leq \frac{C}{\lambda^\kappa}.$$

Thus, from the above inequality, we finally have

$$(\phi(t), \phi(t))^{1/2} + \left(\int_t^T \|\phi(\tau)\|_{1,w}^2 d\tau \right)^{1/2} \leq \frac{C}{\lambda^{\kappa/2}}$$

for all $t \in [0, T)$. Thus, we have proved (3.4.7). $\qquad\square$

Using Lemma 3.4.1, we are able to prove the following convergence result.

Theorem 3.4.3 *Let $\frac{\partial u}{\partial t}$ be sufficiently smooth and the assumptions in Lemma 3.4.1 be fulfilled. The solutions u and u_λ to Problems 3.4.1 and 3.4.2 respectively satisfy*

$$\|u - u_\lambda\|_{L^\infty(0,T;L^2(\Omega))} + \|u - u_\lambda\|_{L^2(0,T;H_{0,w}^1(\Omega))} \leq \frac{C}{\lambda^{\kappa/2}},$$

where C is a positive constant, independent of u, u_λ and λ.

Proof Note $\phi := [u_\lambda - u^*]_+ \in H_{0,w}^1(\Omega)$. We decompose $u - u_\lambda$ as

$$u - u_\lambda = u - u^* - (u_\lambda - u^*) = u - u^* + [u_\lambda - u^*]_- - [u_\lambda - u^*]_+ =: r_\lambda - \phi,$$
(3.4.18)

where $[u_\lambda - u^*]_- = -\min\{u_\lambda - u^*, 0\}$ and $r_\lambda = u - u^* + [u_\lambda - u^*]_-$. From the definition of ϕ, we see $(\phi^\alpha, [u_\lambda - u^*]_-) = [u - u^*]_+^\alpha [u_\lambda - u^*]_- \equiv 0$ for $\alpha > 0$.

It is also easy to show that $r_\lambda = 0$ on $\partial\Omega$. In fact, when $(S, v) \in \partial\Omega$, $u - u^* = u_\lambda - u^* \leq 0$, and so $r_\lambda = u - u^* + [u - u^*]_- = u - u^* - (u - u^*) = 0,$. Thus, $r_\lambda \in H^1_{0,w}(\Omega)$, since all terms in r_λ are in $H^1_w(\Omega)$. Also, from (3.4.18) we see that $u - r_\lambda = u_\lambda - \phi \leq u^*$, since $\phi \geq 0$. Thus, $u - r_\lambda \in \mathcal{K}$.

From (3.4.18) and Lemma 3.4.1, we see that to establish an upper bound for $u - u_\lambda$, we need only to determine an upper bound for r_λ. Choosing $v = u - r_\lambda$ in (3.4.2) and $v = r_\lambda$ in (3.4.5), we have

$$\left(-\frac{\partial u}{\partial t}, -r_\lambda\right) + B(u, -r_\lambda; t) \geq (f, -r_\lambda),$$

$$\left(-\frac{\partial u_\lambda}{\partial t}, r_\lambda\right) + B(u_\lambda, r_\lambda; t) + \lambda(\phi^{1/\kappa}, r_\lambda) = (f, r_\lambda).$$

Adding up the above inequality and equality gives

$$\left(-\frac{\partial(u_\lambda - u)}{\partial t}, r_\lambda\right) + B(u_\lambda - u, r_\lambda; t) + \lambda(\phi^{1/\kappa}, r_\lambda) \geq 0. \tag{3.4.19}$$

Note $\phi \cdot [u_\lambda - u^*]_- = 0$. Using the definition of r_λ, we have

$$(\phi^{1/\kappa}, r_\lambda) = (\phi^{1/\kappa}, u - u^* + [u_\lambda - u^*]_-) = (\phi^{1/\kappa}, u - u^*) \leq 0, \tag{3.4.20}$$

since $\phi \geq 0$ and $u - u^* \leq 0$. Therefore, combining (3.4.19) and (3.4.20) gives

$$\left(-\frac{\partial(u - u_\lambda)}{\partial t}, r_\lambda\right) + B(u - u_\lambda, r_\lambda; t) \leq 0.$$

Using the decomposition of $u - u_\lambda$ in (3.4.18), we have from the above inequality

$$\left(-\frac{\partial r_\lambda}{\partial t}, r_\lambda\right) + B(r_\lambda, r_\lambda; t) \leq \left(-\frac{\partial\phi}{\partial t}, r_\lambda\right) + B(\phi, r_\lambda; t).$$

Note that $r_\lambda(T) = 0$, since u, u^* and u_λ satisfy the payoff condition $g(S)$. Integrating the above from t to T and using (3.4.9) and Cauchy–Schwarz inequality, we obtain

$$\frac{1}{2}(r_\lambda(t), r_\lambda(t)) + \int_t^T B(r_\lambda(\tau), r_\lambda(\tau); \tau)d\tau$$

$$\leq \int_t^T \left(-\frac{\partial\phi(\tau)}{\partial\tau}, r_\lambda(\tau)\right)d\tau + \int_t^T B(\phi(\tau), r_\lambda(\tau); \tau)d\tau$$

$$\leq (\phi(t), r_\lambda(t)) + \int_t^T \left(\phi(\tau), \frac{\partial r_\lambda(\tau)}{\partial\tau}\right)d\tau + \int_t^T B(\phi(\tau), r_\lambda(\tau); \tau)d\tau$$

$$\leq \|\phi\|_{L^\infty(0,T;L^2(I))}\|r_\lambda\|_{L^\infty(0,T;L^2(\Omega))} + C\|\phi\|_{L^2(0,T;H^1_{0,w}(\Omega))}\|r_\lambda\|_{L^2(0,T;H^1_{0,w}(\Omega))}$$

$$+ \int_t^T \left(\phi(\tau), \frac{\partial r_\lambda(\tau)}{\partial t} \right) d\tau \quad t \in (0, T). \tag{3.4.21}$$

Using (2.1.8) and the definition of r_λ we have

$$\phi(\tau) \frac{\partial r_\lambda(\tau)}{\partial \tau} = \phi(\tau) \left(\frac{\partial u(\tau)}{\partial \tau} - \frac{\partial u^*(\tau)}{\partial \tau} \right) = \phi(\tau) \frac{\partial u(\tau)}{\partial \tau} - \phi(\tau) \beta e^{\beta \tau} (V_0 - V^*).$$

Thus, using (3.4.6), we obtain

$$\int_t^T \left(\phi(\tau), \frac{\partial r_\lambda(\tau)}{\partial \tau} \right) d\tau = \int_t^T \left(\phi(\tau), \frac{\partial u(\tau)}{\partial \tau} \right) d\tau - \beta \int_t^T e^{\beta \tau} \left(\phi(\tau), V_0 - V^* \right) d\tau$$

$$\leq C \|\phi\|_{L^p(\Omega)} \left(\left\| \frac{\partial u}{\partial t} \right\|_{L^q(\Omega)} + \|V_0 - V^*\|_{L^q(\Omega)} \right) \leq \frac{C}{\lambda^\kappa},$$

where $p = 1 + 1/\kappa$ and $q = \kappa + 1$. Combining the above upper bound with (3.4.21) and using (3.4.7), we obtain

$$\frac{1}{2}(r_\lambda(t), r_\lambda(t)) + \int_t^T B(r_\lambda(\tau), r_\lambda(\tau); \tau) d\tau$$

$$\leq \|\phi\|_{L^\infty(0,T;L^2(\Omega))} \|r_\lambda\|_{L^\infty(0,T;L^2(\Omega))} + C\|\phi\|_{L^2(0,T;H^1_{0,w}(\Omega))} \|r_\lambda\|_{L^2(0,T;H^1_{0,w}(\Omega))} + \frac{C}{\lambda^\kappa}$$

$$\leq C \left(\|\phi\|_{L^\infty(0,T;L^2(\Omega))} + \|\phi\|_{L^2(0,T;H^1_{0,w}(\Omega))} \right) \left(\|r_\lambda\|_{L^\infty(0,T;L^2(\Omega))} + \|r_\lambda\|_{L^2(0,T;H^1_{0,w}(\Omega))} \right)$$

$$+ C\lambda^{-\kappa} \leq C \left[\lambda^{-k/2} \left(\|r_\lambda\|_{L^\infty(0,T;L^2(\Omega))} + \|r_\lambda\|_{L^2(0,T;H^1_{0,w}(\Omega))} \right) + \lambda^{-k} \right].$$

On the other hand, from (3.1.17) and the definition of $\| \cdot \|_{L^p(0,T;H)}$, we have from the above estimate,

$$\left(\|r_\lambda\|_{L^\infty(0,T;L^2(\Omega))} + \|r_\lambda\|_{L^2(0,T;H^1_{0,w}(\Omega))} \right)^2$$

$$\leq C \left(\frac{1}{2} \|r_\lambda\|^2_{L^\infty(0,T;L^2(\Omega))} + \|r_\lambda\|^2_{L^2(0,T;H^1_{0,w}(\Omega))} \right)$$

$$\leq \frac{1}{2}(r_\lambda(t), r_\lambda(t)) + \int_t^T B(r_\lambda(\tau), r_\lambda(\tau); \tau) d\tau$$

$$\leq C \left[\lambda^{-\kappa/2} \left(\|r_\lambda\|_{L^\infty(0,T;L^2(\Omega))} + \|r_\lambda\|_{L^2(0,T;H^1_{0,w}(\Omega))} \right) + \lambda^{-\kappa} \right]. \tag{3.4.22}$$

This is of the form $y^2 \leq C\rho^{1/2} y + C\rho$ with $\rho = \lambda^{-\kappa}$, which can be rewritten as $(y - \frac{1}{2}Cs\rho^{1/2})^2 \leq \left(C + \frac{C^2}{4} \right) \rho$. Rearrange this inequality gives $y \leq C\rho^{1/2}$. (Recall that C is a generic positive constant.) Thus, applying this analysis to (3.4.22) yields

$$\|r_\lambda\|_{L^\infty(0,T;L^2(\Omega))} + \|r_\lambda\|_{L^2(0,T;H^1_0(\Omega))} \leq \frac{C}{\lambda^{\kappa/2}}.$$

Using the triangle inequality, (3.4.7) and the above inequality, we have from (2.2.17),

$$\|u - u_\lambda\|_{L^\infty(0,T;L^2(\Omega))} + \|u - u_\lambda\|_{L^2(0,T;H^1_{0,w}(\Omega))} \leq$$

$$\|r_\lambda\|_{L^\infty(0,T;L^2(\Omega))} + \|r_\lambda\|_{L^2(0,T;H^1_{0,w}(\Omega))} + \|\phi\|_{L^\infty(0,T;L^2(\Omega))} + \|\phi\|_{L^2(0,T;H^1_{0,w}(\Omega))}$$

$$\leq \frac{C}{\lambda^{\kappa/2}}.$$

Thus, we have proved (2.2.16).   □

To conclude this section, we comment that (3.4.3)–(3.4.4) can be solved numerically by a combination of the FVM in Sect. 3.2 and the damped Newton algorithm in Sect. 2.3.2.

3.5 A Numerical Example

Consider the following test problem.

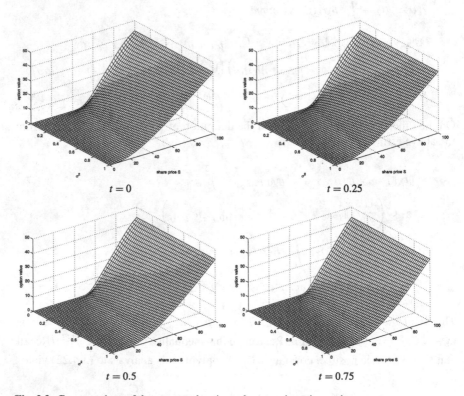

Fig. 3.3 Cross-sections of the computed option value at various time points

Example 3.1 A European call option governed by (3.1.9)–(3.1.11) with $g_3(S) = $
$\max\{0, S - K\}$, $g_1(t) = 0$, and $g_2(t) = S_{max} - K$. The parameters are $S_{max} = 100$,
$\nu_{min} = 0.01$, $\nu_{max} = 1$, $T = 1$, $r = 0.1$, $\rho = 0.9$, $\sigma_\nu = 1$, $\mu_\nu = 0$ and $K = 50$.

We choose a uniform partition (3.2.1) with $N_S = 50$ and $N_\nu = 50$, and $(0, T)$ is
also partitioned uniformly into 50 sub-intervals. The boundary conditions $G_1(S, t)$
and $G_2(S, t)$ in (3.1.11) are determined numerically by solving the constant volatility
counterpart of this example using the 1D FVM in Sect. 1.3 on the aforementioned
uniform mesh. The cross-sections of the numerical solution to Example 3.1 at various
time points are depicted in Fig. 3.3.

References

1. Cox J (1975) Notes on option pricing I: constant elasticity of diffusions. Unpublished draft, Stanford University
2. Garman M (1976) A general theory of asset valuation under diffusion state processes. Working Paper No. 50, University of California, Berkeley
3. Haslinger J, Miettinen M (1999) Finite element method for hemivariational inequalities. Kluwer Academic Publisher, Dordrecht
4. Hull J (2015) Options, future, and other derivatives, 9th edn. Pearson, Boston
5. Hull J, White A (1987) The pricing of options on assets with stochastic volatilities. J Financ 42:281–300
6. Heston SI (1993) A closed-form solution for options with stochastic volatility with applications to bond and currency options. Rev Financ Stud 6:327–343
7. Huang C-S, Hung C-H, Wang S (2006) A fitted finite volume method for the valuation of options on assets with stochastic volatilities. Computing 77:297–320
8. Huang C-S, Hung C-H, Wang S (2010) On convergence of a fitted finite-volume method for the valuation of options on assets with stochastic volatilities. IMA J Numer Anal 30:1101–1120
9. Kufner A (1985) Weighted sobolev spaces. Wiley Inc, New York
10. Varga RS (1962) Matrix iterative analysis. Prentice-Hall, Englewood Cliffs

Chapter 4
Options on One Asset Revisited

Abstract In this chapter we propose a superconvergent Finite Volume Method (FVM) based on that in Sect. 1.3 for the nonlinear penalized Black–Scholes equation governing the valuation of European and American options on one asset. Unlike the FVM in Sect. 1.3, we construct an un-symmetric dual mesh using a set of judiciously chosen points. We show that the approximate flux at these points has a 2nd-order truncation error, instead of the 1st-order one in the FVM in Chap. 1. Thus, the resulting FVM has a higher order accuracy at these points, which are called superconvergent points. Numerical results are presented to demonstrate our theoretical findings.

Keywords Option pricing · Finite volume method · Superconvergence · Truncation error

4.1 The Unsymmetric Finite Volume Method

We revisit the FVM in Sect. 1.3 for (2.2.1). Omitting the subscript λ we rewrite (2.2.1) as the following form:

$$-\frac{\partial u(S, t)}{\partial t} - \frac{\partial[S\rho(u(S, t))]}{\partial S} + c(t)u(S, t) + \varphi(S, t, u(S, t)) = f(S, t) \quad (4.1.1)$$

for $(S, t) \in I \times [0, T)$ satisfying appropriate payoff and boundary conditions as in (2.2.2), where $\rho(u)$ is the flux defined in (1.3.2) and $\varphi(u) = \lambda[u - u^*]_+^{1/k}$ with u^* given in (2.1.8).

We divide $I = (0, S_{\max})$ into N sub-intervals $I_i := (S_i, S_{i+1}), i = 0, 1, \ldots, M - 1$, satisfying $0 = S_0 < S_1 < \cdots < S_N = N$. Let $h_i = S_{i+1} - S_i$ and $h = \max_{0 \leq i \leq M-1} h_i$. Unlike the dual mesh in Sect. 1.3.1 consisting of the mid-points of $\{I_i\}$, we define a general dual mesh with nodes $S_{i-\frac{1}{2}}, i = 0, 1, \ldots, N + 1$, such that $S_{-1/2} = S_0 < S_{\frac{1}{2}} < S_1 < S_{1+\frac{1}{2}} < \cdots < S_{N-1} < S_{N-\frac{1}{2}} < S_N = S_{N+\frac{1}{2}}$. The node $S_{i+\frac{1}{2}}$ is to be determined later for each feasible i.

Integrating (4.1.1) over $(S_{i-\frac{1}{2}}, S_{i+\frac{1}{2}})$, we have, for $i = 1, 2, \ldots, N - 1$,

© The Author(s), under exclusive license to Springer Nature Singapore Pte Ltd. 2020   85
S. Wang, *The Fitted Finite Volume and Power Penalty Methods*
for Option Pricing, SpringerBriefs in Mathematical Methods,
https://doi.org/10.1007/978-981-15-9558-5_4

$$-\int_{S_{i-\frac{1}{2}}}^{S_{i+\frac{1}{2}}} \frac{\partial u}{\partial S} dS - [S\rho(u)]_{S_{i-\frac{1}{2}}}^{S_{i+\frac{1}{2}}} + \int_{S_{i-\frac{1}{2}}}^{S_{i+\frac{1}{2}}} (cu + \varphi(u))dS = \int_{S_{i-\frac{1}{2}}}^{S_{i+\frac{1}{2}}} f(S)dS.$$

Using the one-point quadrature rule, we approximate the above equation by

$$-\frac{\partial u_i}{\partial t} l_i - \left[S_{i+\frac{1}{2}}\rho(u)|_{S_{i+\frac{1}{2}}} - S_{i-1/2}\rho(u)|_{S_{i-\frac{1}{2}}} \right] + (c_i u_i + \varphi_i(u_i))l_i = f_i l_i \quad (4.1.2)$$

for $i = 1, 2, \ldots, N - 1$, where $l_i = S_{i+\frac{1}{2}} - S_{i-\frac{1}{2}}$, $c_i = c(S_i, t)$, $\varphi_i(u_i) = \varphi(S_i, t, u_i)$, $f_i = f(S_i)$, and u_i is an approximation to $u(x_i, t)$ to be determined.

Following the analysis in Sect. 1.3.1, we consider the following two cases.
Case I. $i \geq 1$. We consider the following 2-point boundary value problem:

$$\left(aSw' + b_{i+1/2}w\right)' = 0, \ x \in I_i, \ w(x_i) = u_i, \ w(x_{i+1}) = u_{i+1}. \quad (4.1.3)$$

Let $\alpha_i = b_{i+\frac{1}{2}}/a$. Solving (4.1.3) analytically yields, for $S \in I_i$,

$$\rho_i := b_{i+1/2} \frac{x_{i+1}^{\alpha_i} u_{i+1} - x_i^{\alpha_i} u_i}{x_{i+1}^{\alpha_i} - x_i^{\alpha_i}}, \quad w = \frac{\rho_i}{b_{i+\frac{1}{2}}} - \frac{u_{i+1} - u_i}{S_{i+1}^{\alpha_i} - S_i^{\alpha_i}} (S_i S_{i+1})^{\alpha_i} S^{-\alpha_i}. \quad (4.1.4)$$

Case II. $i = 0$. On I_0, (4.1.3) becomes degenerate at $S = 0$, and (4.1.3) is not uniquely solvable. We look into the asymptotic behaviour of ρ_0 as $S_0 \to 0^+$.

When $\alpha_0 < 0$, we rewrite (4.1.4), for any $S_0 > 0$, as $\rho_0 = b_{\frac{1}{2}} \frac{S_0^{-\alpha_0} S_1^{\alpha_0} u_1 - u_0}{S_0^{-\alpha_0} S_1^{\alpha_0} - 1}$. Taking the limit, we have $\lim_{S_0 \to 0^+} \rho_0 = b_{\frac{1}{2}} u_0$. Similarly, when $\alpha_0 > 0$, from (4.1.4) we have $\lim_{S_0 \to 0^+} \rho_0 = b_{\frac{1}{2}} u_1$. Combining these two situations we obtain

$$\rho_0 = b_{\frac{1}{2}} \frac{1 - \text{sign}(b_{\frac{1}{2}})}{2} u_0 + b_{\frac{1}{2}} \frac{1 + \text{sign}(b_{\frac{1}{2}})}{2} u_1, \quad (4.1.5)$$

since $\alpha_0 = b_{\frac{1}{2}}/a$ and $b_{\frac{1}{2}}$ have the same sign pattern.

Replacing the fluxes in (4.1.2) with their respective approximations defined in (4.1.4) and (4.1.5), we have the following semi-discretized system:

$$-\frac{\partial u_i}{\partial t} l_i + e_{i,i-1} u_{i-1} + e_{i,i} u_i + e_{i,i+1} u_{i+1} + \varphi_i(u_i)l_i = 0, \quad (4.1.6)$$

where

$$e_{1,0} = -\frac{S_1}{2} b_{\frac{1}{2}} \frac{1 - \text{sign}(b_{\frac{1}{2}})}{2}, \quad e_{1,2} = -\frac{b_{1+\frac{1}{2}} S_{1+\frac{1}{2}} S_2^{\alpha_1}}{S_2^{\alpha_1} - S_1^{\alpha_1}}, \quad (4.1.7)$$

$$e_{1,1} = \frac{S_1}{2} b_{\frac{1}{2}} \frac{1 + \text{sign}(b_{\frac{1}{2}})}{2} + \frac{b_{1+\frac{1}{2}} S_{1+\frac{1}{2}} S_1^{\alpha_1}}{S_2^{\alpha_1} - S_1^{\alpha_1}} + c_1 l_1, \quad (4.1.8)$$

$$e_{i,i-1} = -\frac{b_{i-\frac{1}{2}}S_{i-\frac{1}{2}}S_{i-1}^{\alpha_{i-1}}}{S_i^{\alpha_{i-1}} - S_{i-1}^{\alpha_{i-1}}}, \quad e_{i,i+1} = -\frac{b_{i+\frac{1}{2}}S_{i+\frac{1}{2}}S_{i+1}^{\alpha_i}}{S_{i+1}^{\alpha_i} - S_i^{\alpha_i}}, \tag{4.1.9}$$

$$e_{i,i} = \frac{b_{i-\frac{1}{2}}S_{i-\frac{1}{2}}S_i^{\alpha_{i-1}}}{S_i^{\alpha_{i-1}} - S_{i-1}^{\alpha_{i-1}}} + \frac{b_{i+\frac{1}{2}}S_{i+\frac{1}{2}}S_i^{\alpha_i}}{S_{i+1}^{\alpha_i} - S_i^{\alpha_i}} + c_i l_i, \tag{4.1.10}$$

for $i = 1, 2, 3, \ldots, N - 1$. Thus, (4.1.6) is a tri-diagonal nonlinear system in $\boldsymbol{u}(t) := (u_1(t), u_2(t), \ldots, u_{M-1}(t))^\top$ if we take into consideration of $u_0(t) = 0 = u_N(t)$. We write (4.1.6) as the following matrix form:

$$-\frac{\partial u_i(t)}{\partial t}l_i + E_i(t)\boldsymbol{u}(t) + \varphi_i(u_i(t))l_i = f_i(t), \tag{4.1.11}$$

for $i = 1, 2, \ldots, N - 1$, where

$$E_1 = (e_{11}(t), e_{12}(t), 0, \ldots, 0), \quad E_{N-1} = (0, \ldots, 0, e_{N-1,N}(t), e_{N-1,N-1}(t)),$$
$$E_i = (0, .., 0, e_{i,i-1}(t), e_{i,i}(t), e_{i,i+1}(t), 0, \ldots, 0), \quad i = 2, 3, \ldots, N - 2,$$

with $e_{i,j}$'s defined in (4.1.7)–(4.1.10).

For a given positive integer K, let $\{t_k\}_{k=0}^K$ be a set of points in $[0, T]$ satisfying $T = t_0 > t_1 > \cdots > t_K = 0$. On this time-partition, we apply the implicit time-stepping method with a splitting parameter $\theta \in [1/2, 1]$ to (4.1.11) to yield

$$\frac{u_i^{k+1} - u_i^k}{-\Delta t_k}l_i + \theta[E_i^{k+1}\boldsymbol{u}^{k+1} + \varphi_i(u_i^{k+1})l_i] + (1-\theta)[E_i^k\boldsymbol{u}^k + \varphi_i(u_i^k)l_i] = (\theta f_i^{k+1} + (1-\theta)f_i^k)l_i$$

for $k = 0, 1, \ldots, K - 1$ and $i = 1, 2, \ldots, N - 1$ with the terminal condition $\boldsymbol{u}^0 = (u^*(S_1), u^*(S_2), \ldots, u^*(S_{N-1}))^\top$, where $\Delta t_k = t_{k+1} - t_k < 0$, $E_i^k = E_i(t_k)$, $f_i^k = f_i(t_k)$, and $\boldsymbol{u}^k$ denotes the approximation to $\boldsymbol{u}(t_k)$. The above discrete equation can also be rewritten as the following matrix form:

$$(G^k + \theta E^{k+1})\boldsymbol{u}^{k+1} + \theta\Phi(\boldsymbol{u}^{k+1}) = \theta f^{k+1} + (1-\theta)f^k + [G^k - (1-\theta)E^k]\boldsymbol{u}^k$$
$$- (1-\theta)\Phi(\boldsymbol{u}^k), \quad k = 0, 1, \ldots, N - 1. \tag{4.1.12}$$

When $\theta = 1/2$, the time-stepping scheme becomes that of the Crank–Nicolson and when $\theta = 1$, it is the backward Euler scheme. Both of the two cases are unconditionally stable, and they are second and first order accurate in $|\Delta t_k|$, respectively. We show in the following theorem that, when $|\Delta t_k|$ is sufficiently small, $G^k + \theta E^{k+1}$ in (4.1.12) is an M-matrix.

Theorem 4.1.1 *For any given $k = 1, 2, \ldots, K - 1$, if $|\Delta t_k|$ is sufficiently small, then $G^k + \theta E^{k+1}$ in (4.1.12) is a positive-definite M-matrix.*

Proof The proof is identical to that of Theorem 1.3.1. The positive-definiteness of $G^k + \theta E^{k+1}$ is obvious when $|\Delta t_k|$ sufficiently small. This is because $G_k = \frac{1}{|\Delta t_k|}I$, where I denotes the identity matrix. $\square$

4.2 Determination of Superconvergent Points

In what follows, we suppress the time variable t. The discussion below follows that
[1, 2]. for any feasible i, we expand b on I_i as $b(S) = p_i + q_i S + \mathcal{O}(h_i)$, where p_i
and $q_i c$ satisfy $b(S_i) = p_i + q_i S_i$ and $q_i = b'(S_i)$. Assume $\rho(u)$ is given by

$$\left(aSu' + (p_1 + q_i S)u\right)' = g(S), \quad S \in I_i, \tag{4.2.1}$$

where $g(S)$ is unknown. We assume g is continuously differentiable and consider
the following interpolation problem: find $w(S)$ such that

$$\left(aSw' + (p_i + q_i S)w\right)' = 0, \quad S \in I_i, \quad w(S_i) = u(S_i), \quad w(S_{i+1}) = u(S_{i+1}). \tag{4.2.2}$$

Let $v := u - w$. The difference between (4.2.2) and (4.2.1) is

$$\left(aSv' + (p_i + q_i S)v\right)' = g(S), \quad S \in I_i, \quad v(S_i) = 0 = v(S_{i+1}). \tag{4.2.3}$$

Integrating the equation in (4.2.3) and dividing the resulting equation by aS yield

$$v' + \left(\frac{\bar{p}_i}{S} + \bar{q}_i\right) v = \frac{1}{aS}(G(S) + C_1), \tag{4.2.4}$$

where $G = \int g(S)dS$, C_1 is an additive constant, $\bar{p}_i = p_i/a$ and $\bar{q}_i = q_i/a$. The
integrating factor of the above equation is $\mu(S) = e^{\int(\frac{\bar{p}_i}{S}+\bar{q})dS} = S^{\bar{p}_i} e^{\bar{q}_i S}$, and thus
the general solution to (4.2.4) is

$$v(S) = S^{-\alpha_i} e^{-\bar{q}_i S} \left(\int_{S_i}^{S} t^{\bar{p}_i} e^{\bar{q}_i t} \frac{G(t) + C_1}{at} dt + C_2\right) dt, \tag{4.2.5}$$

where C_2 is also an additive constant. Using the boundary conditions in (4.2.3), we
have from (4.2.5) $C_2 = 0 = \int_{S_i}^{S_{i+1}} t^{\bar{p}_i} e^{\bar{q}_i t} \frac{G(t)+C_1}{at} dt$, from which we obtain

$$\int_{S_i}^{S_{i+1}} t^{\alpha_i - 1} e^{\bar{q}_i t} G(t)dt = -C_1 \int_{S_i}^{S_{i+1}} t^{\bar{p}_i - 1} e^{\bar{q}_i t} dt =: -C_1 Q(S_i, S_{i+1}).$$

Thus, $C_1 = -Q^{-1}(S_i, S_{i+1}) \int_{S_i}^{S_{i+1}} t^{\bar{p}_i-1} e^{\bar{q}_i t} G(t)dt$, and so, we can write $G(S) - C_1$
as

$$G(S) + C_1 = \frac{1}{Q(S_i, S_{i+1})} \int_{S_i}^{S_{i+1}} (G(S) - G(t)) t^{\alpha_i - 1} e^{\bar{q}_i t} dt \tag{4.2.6}$$

for $S \in I_i$. Replacing $G(t) - G(S)$ in (4.2.6) with its Taylor expansion $G(t) -
G(S) = g(S)(t - x) + g'(S)(t - S)^2 + \cdots$, we have

$$G(S) + C_1 = -\frac{1}{Q(S_i, S_{i+1})} \int_{S_i}^{S_{i+1}} t^{\bar{p}_i - 1} e^{\bar{q}_i t} \left[g(S)(t - S) + g'(x)(t - S)^2 + \cdots \right] dt$$

$$= -\frac{g(S)}{Q(S_i, S_{i+1})} \int_{S_i}^{S_{i+1}} t^{\bar{p}_i - 1} e^{\bar{q}_i t} (t - S) dt + g'(S) \mathcal{O}(h_i^2)$$

$$=: -\frac{g(S)}{Q(S_i, S_{i+1})} [R(S_i, S_{i+1}) - S Q(S_i, S_{i+1})] + g'(S) \mathcal{O}(h_i^2) \qquad (4.2.7)$$

with $R(S_i, S_{i+1}) = \int_{S_i}^{S_{i+1}} t^{\bar{p}_i} e^{\bar{q}_i t} dt$. Therefore, if we choose

$$S_{i+\frac{1}{2}} := \frac{R(S_i, S_{i+1})}{Q(S_i, S_{i+1})}, \qquad (4.2.8)$$

then the first term on the RHS of (4.2.7) becomes zero. In this case, from (4.2.4) and (4.2.7) we see that

$$[aSv' + b_{i+\frac{1}{2}}v]_{S_{i+\frac{1}{2}}} = [aSv' + (p_i + q_i S)v]_{S_{i+\frac{1}{2}}} + [(S_{i+\frac{1}{2}} - (p_i + q_i S))v]_{S_{i+\frac{1}{2}}}$$

$$= G(S_{i+\frac{1}{2}}) + C_1 + b''(S_i)\mathcal{O}(h_i^2) = g'(S_{i+\frac{1}{2}})\mathcal{O}(h_i^2) + C_1 + b''(S_i)\mathcal{O}(h_i^2), \quad (4.2.9)$$

where $b''(S_i)\mathcal{O}(h_i^2)$ is the remainder of the Taylor's expansion.

In computation, $Q(S_i, S_{i+1})$ and $R(S_i, S_{i+1})$ can be calculated using an appropriate numerical quadrature rule. However, using the expansion $e^{\bar{q}_i t} = e^{\bar{q}_i S_i}[1 + \bar{q}_i(t - S_i)] + \mathcal{O}(h_i^2)$, we may represent $Q(S_i, S_{i+1})$ and $R(S_i, S_{i+1})$ as follows.

$$Q(S_i, S_{i+1}) = e^{\bar{q}_i S_i} \left[\frac{1 - \bar{q}_i S_i}{\bar{p}_i} (S_{i+1}^{\bar{p}_i} - S_i^{\bar{p}_i}) + \frac{\bar{q}_i}{\bar{p}_i + 1}(S_{i+1}^{\bar{p}_i + 1} - S_i^{\bar{p}_i + 1}) \right] + \mathcal{O}(h_i^3),$$

$$R(S_i, S_{i+1}) = e^{\bar{q}_i S_i} \left[\frac{1 - \bar{q}_i S_i}{\bar{p}_i + 1} (S_{i+1}^{\bar{p}_i + 1} - S_i^{\bar{p}_i + 1}) + \frac{\bar{q}_i}{\bar{p}_i + 2}(S_{i+1}^{\bar{p}_i + 2} - S_i^{\bar{p}_i + 2}) \right] + \mathcal{O}(h_i^3).$$

Using these and (4.2.8), we rewrite $S_{i+\frac{1}{2}}$ as

$$S_{i+\frac{1}{2}} = \frac{\frac{1 - \bar{q}_i S_i}{\bar{p}_i + 1}(S_{i+1}^{\bar{p}_i + 1} - S_i^{\bar{p}_i + 1}) + \frac{\bar{q}_i}{\bar{p}_i + 2}(S_{i+1}^{\bar{p}_i + 2} - S_i^{\bar{p}_i + 2})}{\frac{1 - \bar{q}_i S_i}{\bar{p}_i}(S_{i+1}^{\bar{p}_i} - S_i^{\bar{p}_i}) + \frac{\bar{q}_i}{\bar{p}_i + 1}(S_{i+1}^{\bar{p}_i + 1} - S_i^{\bar{p}_i + 1})} + \mathcal{O}(h_i^2). \qquad (4.2.10)$$

Equation (4.2.10) provides a breakpoint for constructing the dual mesh if we omit $\mathcal{O}(h_i^2)$, which may also depends on t. At this point, the approximate flux has a 2nd-order truncation error.

4.3 Superconvergent Points When b Is Independent of S

When b is constant, $p_i = b$ and $q_i = 0$, Thus, (4.2.10) becomes

$$S_{i+\frac{1}{2}} := \frac{\alpha}{\alpha+1} \frac{S_{i+1}^{\alpha+1} - S_i^{\alpha+1}}{S_{i+1}^{\alpha} - S_i^{\alpha}}, \quad i = 1, 2, \ldots, N-1, \tag{4.3.1}$$

when $\alpha := \frac{b}{a} \neq 0, -1$. If $\alpha = 0$ or -1, using the 2nd integral in (4.2.7), we have

$$\int_{S_i}^{S_{i+1}} t^{\alpha-1}(t-S)dt = \begin{cases} S_{i+1} - S_i - x \ln \frac{S_{i+1}}{S_i}, & \alpha = 0, \\ \ln \frac{S_{i+1}}{S_i} - \frac{x}{S_i S_{i+1}}(S_{i+1} - S_i), & \alpha = -1. \end{cases}$$

Thus, we have

$$S_{i+\frac{1}{2}} = \begin{cases} S_i S_{i+1}(\ln \frac{S_{i+1}}{S_i})/h_i, & \alpha = -1, \\ h_i / \ln \frac{S_{i+1}}{S_i}, & \alpha = 0 \end{cases} \quad i = 1, 2, \ldots, N-1. \tag{4.3.2}$$

When h_i is sufficiently small, using a symbolic computation package such as Maple or Mathematica, we can expand $S_{i+\frac{1}{2}}$ as the following Taylor polynomial:

$$S_{i+\frac{1}{2}} = S_i + \frac{1}{2}h_i + \frac{\alpha-1}{12}\frac{h_i^2}{S_i} - \frac{\alpha-1}{24}\frac{h_i^3}{S_i^2} + \mathcal{O}\left(\frac{h_i^4}{S_i^3}\right), \quad i = 1, 2, \ldots, N-1. \tag{4.3.3}$$

From (4.3.3), we see that $S_{i+\frac{1}{2}}$ an increasing function of α_i. Also, using (4.3.1), it is easy to verify that $\lim_{\alpha_i \to -\infty} S_{i+\frac{1}{2}} = S_i$ and $\lim_{\alpha_i \to \infty} S_{i+\frac{1}{2}} = S_{i+1}$. A special case of $S_{i+1}(\alpha)$ with $S_i = 1$ and $S_{i+1} = 2$ is depicted in Fig. 4.1.

Fig. 4.1 The superconvergence point in $(1, 2)$ as a function of α_i

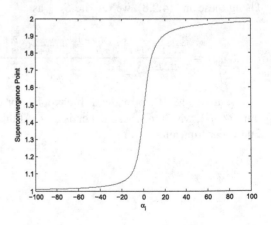

Remark 4.3.1 Note that when $i = 0$, S_0^α (respective $S_0^{\alpha+1}$) is singular when $\alpha < 0$ (respectively $\alpha + 1 < 0$). In this case, we simply choose $S_{\frac{1}{2}} = h_0/2$. This will not affect our overall error if we choose $h_0 = \mathcal{O}(h^2)$.

4.4 Local Error Estimates at the Superconvergent Points

The following theorem establishes error bounds for the interpolated flux and solution.

Theorem 4.4.1 *Let u be the solution to (4.1.1) with the boundary and terminal conditions (2.2.2), and w be the solution to (4.2.2). If u is three times continuously differentiable at $S_{i+\frac{1}{2}}$, then, for any $i = 1, 2, \ldots, N - 1$, we have*

$$\rho_i(u(S_{i+\frac{1}{2}})) - \rho_i(w(S_{i+\frac{1}{2}})) = g'(S_{i+\frac{1}{2}})\mathcal{O}(h_i^2), \tag{4.4.1}$$

$$u(S_{i+\frac{1}{2}}) - w(S_{i+\frac{1}{2}}) = g(S_{i+\frac{1}{2}})\mathcal{O}\left(\frac{\delta_i^2}{S_i}\right) + g'(S_{i+1\frac{1}{2}})\mathcal{O}\left(\frac{h_i^2\delta_i}{S_i}\right), \tag{4.4.2}$$

where $\delta_i = S_{i+\frac{1}{2}} - S_i$ and g is the function defined in (4.2.1).

Proof Since $\rho_i(v) = \rho_i(u) - \rho_i(w)$, (4.2.9) implies (4.4.1), because $b''(S) = 0$.

To prove (4.4.2), we set $S = S_{i+\frac{1}{2}}$ in (4.2.5). Expand $G(t) - C_1$ in the resulting expression as a Taylor's series at $S_{i+\frac{1}{2}}$ with a remainder and using (4.4.1), we have

$$v(S_{i+\frac{1}{2}}) = S_{i+\frac{1}{2}}^{-\alpha} \int_{S_i}^{S_{i+\frac{1}{2}}} \frac{t^{\alpha-1}}{a}\left[(G(S_{i+\frac{1}{2}}) + C_1) + g(S_{i+\frac{1}{2}})(t - S_{i+\frac{1}{2}}) + \mathcal{O}((t - S_{i+\frac{1}{2}})^2)\right] dt$$

$$= S_{i+\frac{1}{2}}^{-\alpha}\left[g'(S_{i+\frac{1}{2}})\mathcal{O}(h_i^2) + g(S_{i+\frac{1}{2}})\mathcal{O}(\delta_i) + \mathcal{O}(\delta_i^2)\right]\int_{S_i}^{S_{i+\frac{1}{2}}} t^{\alpha_i - 1}dt$$

$$= \left[g(S_{i+\frac{1}{2}})\mathcal{O}(\delta_i) + g'(S_{i+\frac{1}{2}})\mathcal{O}(h_i^2)\right]\frac{1 - (S_i/S_{i+\frac{1}{2}})^\alpha}{\alpha}, \tag{4.4.3}$$

where $\delta_i = S_{i+1/2} - S_i$. Using the Taylor's expansion $\frac{1 - (S_i/S_{i+\frac{1}{2}})^\alpha}{\alpha} = \frac{\delta_i}{S_i} + \mathcal{O}(\delta_i^2/S_i^2)$, we have from (4.4.3)

$$u(S_{i+\frac{1}{2}}) - w(S_{i+\frac{1}{2}}) = v(S_{i+\frac{1}{2}}) = g(S_{i+\frac{1}{2}})\mathcal{O}\left(\frac{\delta_i^2}{S_i}\right) + g'(S_{i+\frac{1}{2}})\mathcal{O}\left(\frac{h_i^2\delta_i}{S_i}\right).$$

This is (4.4.2), and we have proved the theorem. $\square$

Using Theorem 4.4.1, we estimate the error between the flux $\rho(u)$ and the interpolant of $\rho_i(w)$ in the following theorem.

Theorem 4.4.2 *Assume that $\rho(u)$ defined in (1.3.2) has 2nd-derivative with respect to u which is bounded on $\bar{I}$. If the conditions in Theorem 4.4.1 are fulfilled, the following estimate holds.*

$$|\rho(u(x)) - \Pi(\rho_i(w(x)))| \leq Ch_i^2$$

for $S \in [S_{i-\frac{1}{2}}, S_{i+\frac{1}{2}}]$, where w and $\rho_i(w)$ are defined in Theorem 4.4.1, C is a positive constant, independent of h_i and Π denotes the usual linear interpolation operator defined on $[S_{i-\frac{1}{2}}, S_{i+\frac{1}{2}}]$ such that $\rho_i(w(S_{i\pm\frac{1}{2}})) = \Pi(\rho_i(w(S_{i\pm\frac{1}{2}})))$.

Proof From Theorem 4.4.1, we have

$$|\rho(u(S)) - \Pi(\rho_i(w(S)))| \leq |\rho(u(S)) - \Pi(\rho_i(u(S)))| + |\Pi\rho(u_i(S)) - \Pi(\rho_i(w(S)))|$$
$$\leq \mathcal{O}(h_i^2) + \chi_{i-\frac{1}{2}}(S)|\rho(u_i(S_{i-\frac{1}{2}})) - (\rho_i(w(S_{i-\frac{1}{2}}))|$$
$$+ \chi_{i+\frac{1}{2}}(S)|\rho(u_i(S_{i+\frac{1}{2}})) - (\rho_i(w(S_{i+\frac{1}{2}}))| \leq \mathcal{O}(h_i^2),$$

where $\chi_{i\pm\frac{1}{2}}(x)$ denote the usual linear basis (hat) functions associate with $S_{i\pm\frac{1}{2}}$ used in interpolation. These functions are in between 0 and 1. Thus, we have proved this theorem.   □

When i is close to 0, $\mathcal{O}(S_i) = \mathcal{O}(h_i)$ so that $\delta_i^2/h_i = \mathcal{O}(h_i)$. If we choose a mesh satisfying $h_i = \mathcal{O}(h)$, then (4.4.2) only provides an $\mathcal{O}(h)$ error for $u(S_{i+\frac{1}{2}}) - w(S_{i+\frac{1}{2}})$. Therefore, a non-uniform or graded mesh needs to be used near $S = 0$ so that $h_i = \mathcal{O}(h^2)$ when i is close to 0.

4.5   Numerical Experiments

We now demonstrate numerically that the points in (4.3.1), (4.2.10) and (4.3.2) are superconvergent points.

Example 4.1 Consider approximating the flux $aSu' + bu = G(S)$ on the interval $I = [S_1, S_2]$ by a constant defined in (4.1.4), where $u(S) = e^S$.

To demonstrate the superconvergence at the points, we consider the following two different sets of coefficients.

Case 1. $a = 0.5$, $b = -1.5$, $S_1 = 1$ and $S_2 = 1.2$.

This is the case that both a and b are constant, and $\rho(u) = e^S(aS + b)$. Using (4.3.1), we find that the superconvergent point $S_{1+1/2} \approx 1.033607$. The exact and approximate fluxes are plotted in Fig. 4.2a in which we also circle the point $(S_{1+1/2}, \rho(u(S_{1+1/2})))$. As predicted, the approximation defined in (4.1.4) achieves a high-order accuracy at $S_{1+1/2}$.

Case 2. $a = 0.5$, $b = 1 + S^2$, $S_1 = 0.2$ and $S_2 = 0.25$.

In this case the exact flux is $\rho(u) = e^S(aS + 1 + S^2)$. Since b is non-constant, we approximate it by $b \approx p + qS$ with $p = 1 - S_i^2$ and $q = 2S_1$. The superconvergent point $S_{1+1/2}$ is approximated by (4.2.10). As in Case 1, we plot the fluxes, as well as $(S_{1+1/2}, \rho(u(S_{1+1/2})))$, in Fig. 4.2b, from which we see that the approximate flux also has a high-order accuracy at $S_{1+1/2}$.

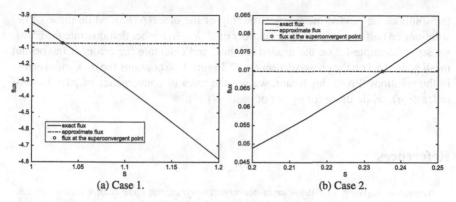

Fig. 4.2 Computed and exact fluxes for Example 4.1 using two different data sets

Example 4.2 We consider pricing the European put option governed by (1.2.3)–(1.2.4) with $K = 50$ and $T = 1$. Other parameters are $\sigma = 0.4, r = 0.03$, and $D = 0$. We choose $S_{\max} = 100$ and $g_1(t) = K \exp(t - T), g_2(t) = 0, g_3(S) = \max\{0, K - S\}$.

We use this example to demonstrate the rates of convergence of the numerical method. For positive integers N and M, we choose two different meshes: (1) uniform mesh with $S_i = i S_{\max}/N$ and $|\Delta t_k| = k/M$; (2) spatially graded mesh $S_i = (i/N)^2 S_{\max}$ and $|\Delta t_k| = k/M$ as in [2], for a feasible pair (i, k). We use the numerical solution from (4.1.12) with $\theta = 1/2$ on the graded mesh with $N = 2560 = M$ as the 'exact' or reference solution. Using this reference solution, we compute the maximum errors at $t = 0$ in u_h and $\rho(u_h)$ on various graded and uniform meshes. Figure 4.3a contains the maximum flux errors in the numerical solutions at the points given in (4.3.1) using the graded and uniform meshes. For comparison, we also plot $y = h^2$ and $y = h$ in Fig. 4.3a with $h = 1/N = 1/M$, from which we see the rates of convergence for

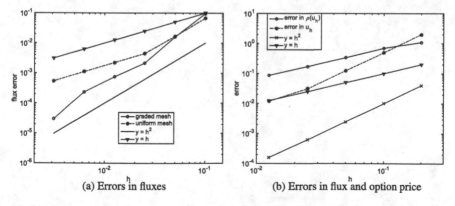

Fig. 4.3 Computed maximum errors for Example 4.2 on different meshes: **a** errors at dual mesh points; **b** errors at primal mesh points

the solutions on graded meshes are roughly of the order $\mathcal{O}(h^2)$, while those of the solutions on uniform meshes have an order of $\mathcal{O}(h)$. To further demonstrate the FVM, we solve Example 4.2 on the graded meshes, and calculate the errors at the primal mesh nodes $\{S_i\}$ on the cross-section $t = 0$. Figure 4.3b contains the maximum errors in the solutions. From this figure, we see the rates of convergence of $\rho(u_h)$ are of order $\mathcal{O}(h)$, while those for u_h are of order $\mathcal{O}(h^2)$.

References

1. Angermann L, Wang S (2019) A super-convergent unsymmetric finite volume method for convection diffusion equations. J Comput Appl Math 358:179–189
2. Wang S, Zhang S, Fang Z (2015) A superconvergent fitted finite volume method for Black Scholes equations governing European and American option valuation. Numer Methods Part Differ Equ 31:1190–1208

Printed in the United States
by Baker & Taylor Publisher Services